University of Pennsylvania

THE ELDRIDGE REEVES JOHNSON FOUNDATION
FOR MEDICAL PHYSICS

Electrical Signs of Nervous Activity

Photograph by *The Evening Bulletin*,
Philadelphia, Pennsylvania

Honorary Degrees Of Doctor of Science
Dr. Herbert Spencer Gasser (left) and Dr. Joseph Erlanger.

Electrical Signs of Nervous Activity

By

Joseph Erlanger

and

Herbert S. Gasser

Philadelphia
UNIVERSITY OF PENNSYLVANIA PRESS

Second Edition

Library of Congress Catalog Card Number: 68-56091

8122-7586-1

Printed in The United States of America

THE JOHNSON FOUNDATION

THE first Johnson Foundation Lectures were delivered in 1930. Their origin was this:

The Eldridge Reeves Johnson Foundation for Medical Physics was established at the University of Pennsylvania in 1929 with a gift of approximately one million dollars from the man for whom it was named.

In an adventurous spirit, Professor Alfred Stengel, Vice President for Medical Affairs, and his colleagues on the medical faculty entrusted the creation of this most highly endowed unit of the University to five young scientists, thirty years of age or less: a physicist-turned-biologist, an engineer, a physician, a psychologist, and a physiologist. I recall saying to Professor Stengel after six months of complete freedom: "Are we developing biophysics and medical physics as you intended?" Said he: "If you knew how unformed were our ideas as to the appropriate scope of biophysics and our hopes for the Foundation, you would feel free to pioneer as you desire. The University will be well satisfied if the research and teachings of the Johnson Foundation emphasize the unity of the physical and biological sciences and wisely define the nature of biophysics. What you do and say in this first university department of biophysics will stimulate and influence the future course of this field of investigation."

As a part of our enthusiastic endeavors to further fruitful cooperation between the physical and biological sciences, we decided to invite some of our most eminent teachers to tell of their scientific adventures in an occasional series of public lectures to be published by the University Press.

In 1930, A. V. Hill, Fulerton Research Professor in The Royal Society, gave the first lectures under the appropriate title "Adventures in Biophysics." He was followed in 1931 by another Fulerton Professor, the now Lord Adrian of Cambridge, whose subject was "The Mechanism of Nervous Action: Electrical Studies of the Neuron." The lectures by Erlanger and Gasser which are reprinted in this volume were delivered in 1936. All of the lecturers had been physiologists who had used physical methods and concepts in their biological research, all were friends who had given much encouragement to the young biophysicists of the Johnson Foundation, all were Nobel laureates and honorary graduates of the University of Pennsylvania.

In 1937 Irving Langmuir, Nobel laureate in physics, told of the fascination biological problems may hold for physicists as he described his research on protein monolayers. Before his lectures were ready for publication, he was distracted by the obligations to war efforts. The series of Johnson Foundation Lectures was thus ended.

In his introduction to the first volume in the series, Professor Hill said: "It was Borelli two hundred and fifty years ago who affirmed that the study of the motion of animals, no less than Astronomy, is a part of physics, to be enlarged and adorned [note the word 'adorned'] by mathematical demonstrations. The Johnson Foundation for Medical Physics, in which I have the honour to inaugurate this series of lectures, is intended to fulfill Borelli's precept. I wish it good fortune in its enviable task: may it adorn, as well as extend, the fields of biological and medical knowledge!"

That challenge has been heeded for well nigh forty years. My colleagues and successors in the Johnson Foundation have certainly extended their chosen fields of study and adorned them.

DETLEV W. BRONK
Director of the Johnson Foundation, 1929-1949

FOREWORD

THE most important reason for reprinting a book is demand. In general, and not unnaturally, need must be expressed for action to be taken, and the expression must be conveyed to those in a position to act. Concerning *Electrical Signs of Nervous Activity* by Joseph Erlanger and Herbert S. Gasser, one in a series of Johnson Foundation Lectures established by Detlev W. Bronk, this need has been expressed a number of times on a number of occasions by a number of people. I have served as the means for conveying their expressions to the University of Pennsylvania Press. The Press has responded graciously, and in so doing does a service to neurophysiology and to neurophysiologists.

The virtue of this book does not now lie in its being the latest word, for there has been amendment since its initial appearance, presaged in the authors' preface and made in no small measure by themselves. To cite but one example, an important one, there remain today in the sequence of elevations described as recordable from mammalian cutaneous nerves only those known as alpha and delta, to represent the myelinated fibers, and the C elevations to represent the unmyelinated fibers. The action of C fibers, so little known in 1937, has since been explored in depth and related in detail. What then was called the C fiber elevation now is a sequence of seven elevations. The fiber spectrum of a cutaneous nerve would appear to be accommodated. But that which now is the latest word may itself in the future be recast in some unknown degree when some new stage of understanding has been reached. Yet the milestones of the past and those of the present are the markers of progress, and a milestone passed is nonetheless of enduring value to those who would follow the path of learning. This book is such a milestone.

DAVID P. C. LLOYD

June, 1968

ACKNOWLEDGMENTS

There are appended bibliographies of the two authors of this book for the reason that bibliographies are useful. I am indebted to Dr. Joseph Hinsey, Mrs. Alfred E. Mirsky, and Professor C. C. Hunt for aid, but I fear there may be errors for which, in the last analysis, I must be responsible.

D.P.C.L.

PREFACE

TWO summers ago, as we sat upon a ledge high up in the Rocky Mountains, resting from our walk and viewing the panorama of lofty peaks spread out before us, our conversation turned to problems of nerve physiology. We were on a holiday together. One of us had just arrived, and much had gone on in our then widely separated laboratories which we had not had an opportunity to discuss. After a while, during a pause in the conversation, one of us said: "But how shall we ever be able to collaborate in a set of lectures when there are so many points which we interpret differently?"

When the time came, however, for the planning of these lectures, the answer to the query was not difficult to find. In accord with our special interests, the subject matter could readily be divided into two parts, to be treated independently without bringing the opinions expressed in one part into conflict with those expressed in the other. And so each of us granted to the other the privilege of telling his part of the story exactly as he saw it, and it was agreed that the statements of each were to remain unchallenged until the discussions were resumed after publication.

We trust that in this arrangement Doctor Bronk will not see any lack of appreciation on our part of the honor which he has conferred upon us in asking us to give this third series of lectures under the auspices of the Eldridge Reeves Johnson Foundation, for indeed our gratitude to him for his invitation is most deeply felt—the more so because of the encouragement he has given us to organize the subject in its present state, rather than to delay longer in the belief that a more complete development would bring a more appropriate time of presentation.

J. E.
H. S. G.

April 7, 1936

CONTENTS

I

THE ANALYSIS OF THE COMPOUND ACTION POTENTIAL OF NERVE

THE second series of lectures under the auspices of this Foundation, delivered in 1931 (Adrian, 1932), carries the title, "The Mechanism of Nervous Action"—the present series, the third, "Electrical Signs of Nervous Activity." The similarity of the two titles bears witness to the influence developments in the fundamental sciences exert upon trends and progress in physiology. For it will become evident that but for contributions made by physicists in the field of vacuum-tube methods, which in turn rest upon experiments still more fundamental, on the theory and the production of electromagnetic waves, neither of these series of lectures would have been possible. Nerve currents may be exceedingly feeble; not only that, they may also be exceedingly brief. To record them, as Adrian's historical treatment of the subject makes clear, has taxed the ingenuity of physiologists through the years. Until vacuum-tube amplification of these currents became possible the available recording methods, though ingenious enough, were in most respects quite inadequate. The progress made by Adrian and his collaborators became possible through the use of amplifiers in conjunction, first, with the Lippmann capillary electrometer (Adrian, 1926) and finally, in 1928, with the electromagnetic recorder as developed by Matthews. The minute nerve currents are amplified to the point where, in the latter case, for instance, they become sufficient to operate the armature of an electromagnet carrying a mirror, and they are visualized by recording photographically the movements of a beam of light reflected from the mirror.

Adaptation of the Electron Oscillograph

In 1921 my then associate at Washington University and present associate on this lectureship had, along with Newcomer, developed an amplifier suited to the purposes of physiology, and those two investigators at that early date had succeeded in building up nerve currents with it to the point where they could be recorded fairly accurately with the Einthoven string galvanometer. But neither the string galvanometer, nor the capillary electrometer, nor even the Matthews oscillograph is quick enough to reproduce true to form the configuration of the briefer of the nerve currents, the quickest of which attain their crests, as we shall see, in perhaps 0.0002 second or even in 0.0001 second, or 0.1 millisecond (msec., as we say). To record accurately any oscillation the recording device should, according to Frank, have a natural frequency at least five times faster than the frequency to be recorded, which in this case then should run up to 50,000 per sec. The string galvanometer, even as improved by Forbes and collaborators (1931), falls short of this requirement; and so, too, does the Matthews oscillograph. In 1890 Braun devised the cathode ray oscillograph, a vacuum tube now popularized through its employment in the visualization of pictures in television, which it does by virtue of its action as a recorder of electrical potentials. In the Braun tube (Fig. 1) the moving part consists of an electron beam which in the present experiments is bent by the action of the amplified potential derived from the nerve, so that the changes in the potential occurring in association with the nerve's action are traced on the screen of the tube where they appear as an illuminated figure, which, if desired, can be photographed. Since the mass of the moving part of this mechanism, namely, the electron, for all practical purposes is negligible, the electron being the smallest particle known, there is scarcely any limit to the rate of potential change the Braun tube can be made to follow. Potentials oscillating at a frequency as high as 200 million per second have been accurately recorded by it when directly impressed upon the deflecting plates of the tube. But changes in current intensity as fast as these would be distorted if passed through an amplifier on their way to the Braun tube. Action potentials, however, are very much slower, so much so that they come through the amplifier without significant distortion and are therefore faithfully reproduced by the recording mechanism. The various objections which from time to time have been raised to the electron oscillograph as a recording instrument for physiological processes have all been met by the improvements which have taken place in tube construction and in accessory methods (see Adrian, 1935). It is possible now to photograph the movements of the spot of light quite as readily as the deflections of any other recording device, and the Braun tube has the advantage of visualizing the pictures—one sees the picture while it is being taken. The improvements in technique will become obvious as the lectures proceed, for the pictures to be shown are reproductions of records that have been made at various stages of progress in method.

The Conducted Action Potential of a Mixed Nerve

After having succeeded in adapting the electron oscillograph to physiological ends the first task set was to obtain a record of the action potential from the nerve which had in a sense served physiologists as a standard in investigations of this type, the sciatic of the (bull)frog. This nerve has the advantage of length and of consisting of many hundreds of fibers of every size and variety. Since Adrian has so recently, and in a Johnson Lecture, described methods in detail it will suffice to remind you cursorily of the essential principles employed in the recording of nerve action potentials. In the present experiments the nerve, after excision, is placed on two pairs of electrodes (see

Fig. 1). One pair (S), close to the central end of the preparation, is connected with the source of current used to stimulate the nerve—say, a break shock from an induction coil. The other pair, the lead electrodes (L), is placed distally, one terminal on the killed end and the second on adjacent but intact surface of the nerve. The action potential traveling along the nerve from the stimulated locus (S) is then recorded as monophasically as is possible (Bishop, Erlanger, and Gasser, 1926) at the

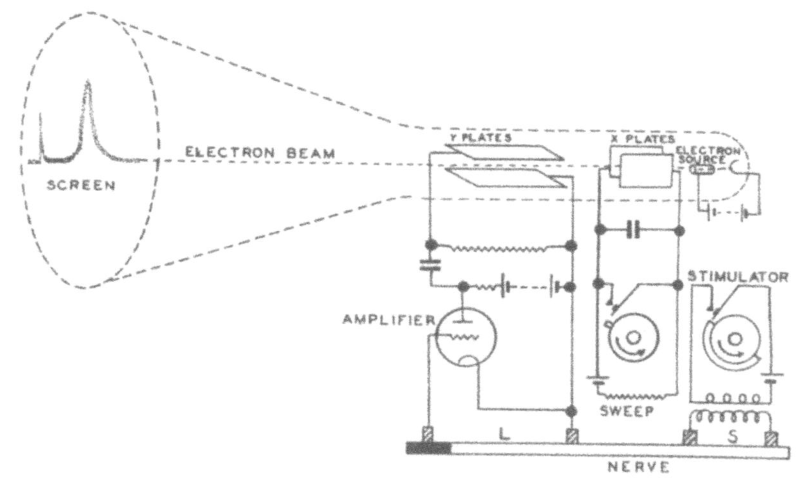

FIG. 1. Simplified diagram of apparatus.[1]

electrode of the distal pair that rests on intact nerve. Fig. 1 shows the circuits in simplified form; the complete wiring of the amplifying and timing or spreading devices is shown in Fig. 2.

The best record of a conducted action potential afforded by the literature down to 1922 is reproduced in Fig. 3. It was made by Garten (1910) with the string galvanometer. Note that the time is recorded in fifths of seconds, and that the figure is simple. The complete, conducted action

[1] I wish here to express my thanks to E. A. Blair for valuable assistance in the preparation of these lectures.

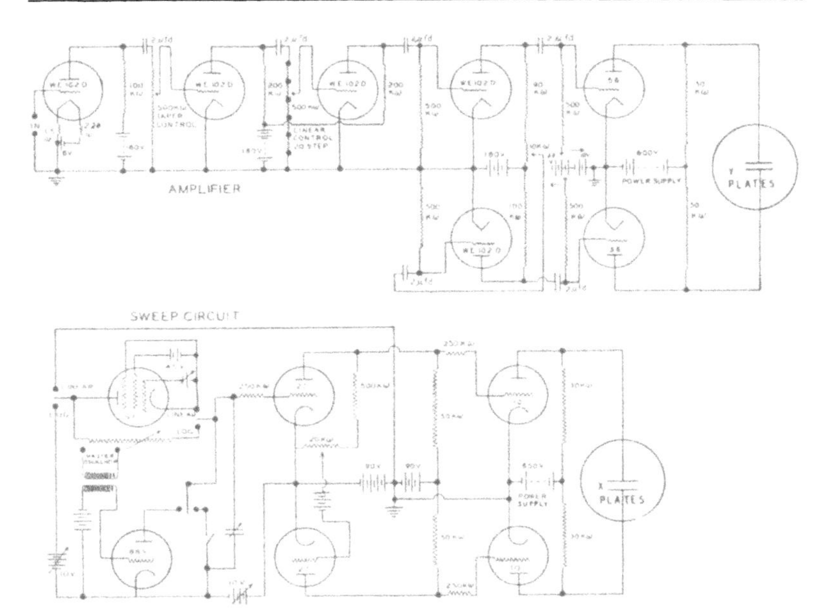

FIG. 2. Amplifying and spreading devices (the latter according to Peugnet).

potential of the frog's sciatic as written by the electron oscillograph can be obtained best by making a series of pictures, one after the other, as the strength of the stimulating current is gradually increased under appropriate

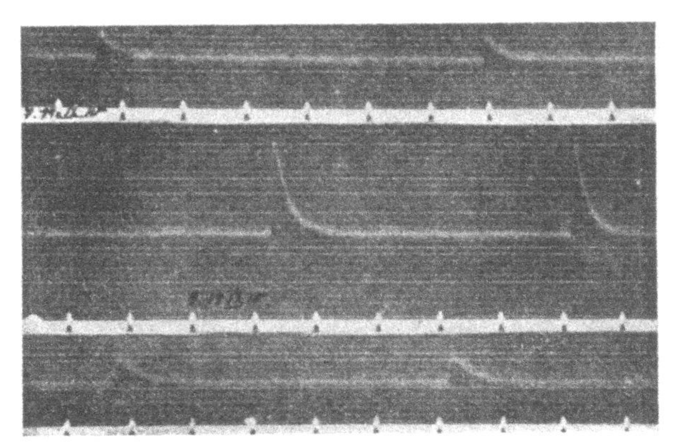

FIG. 3. Action potential of the frog's sciatic recorded with the string galvanometer. Taken from Garten (1910).

recording conditions, to be described in due course. But it will make for clearness, we believe, to present the picture of the action potential in its entirety before discussing the factors which determine its evolution.

The first action potential of the series to be presented (Fig. 4) consists of camera pictures of the deflections made by the most recent technique. It comes from an unusually long nerve; the sciatic trunk was stimulated with a shock strong enough to excite all of the fibers, and the action

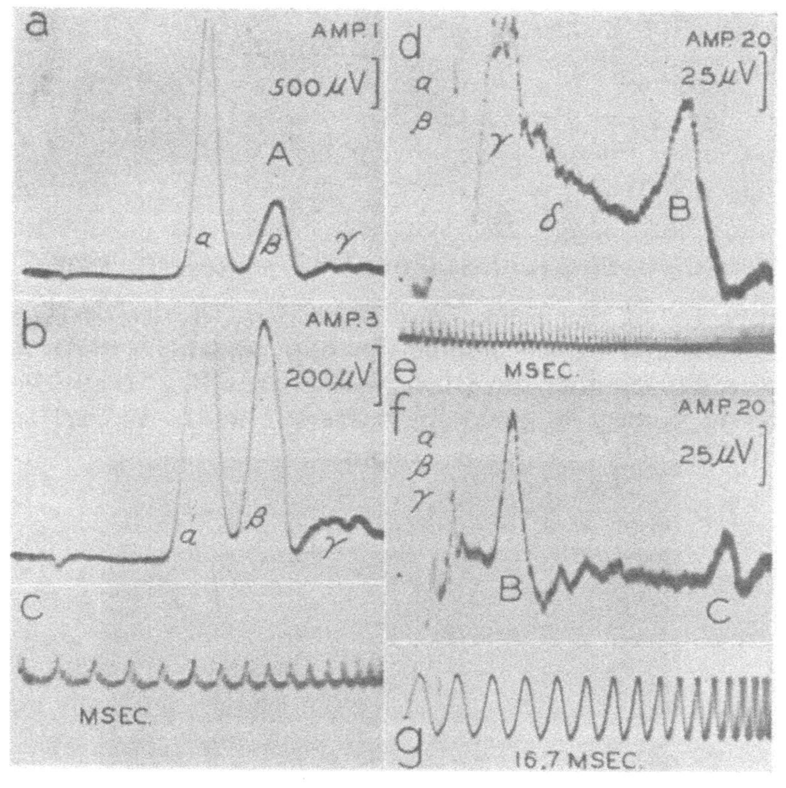

Fig. 4. Complete action potential of bullfrog's peroneal nerve; distance of conduction 13.1 cm. The time falls off logarithmically from left to right; in *c* and *e* the period of the larger oscillations is one millisecond; in *g* the oscillation rate is 60 per sec.; *c* applies to *a* and *b*, *e* to *d* and *g* to *f*. For further description, see text.

potential was led from the peroneal branch, the distance of conduction being 13.1 cm.* Note that time intervals on the records fall off logarithmically from left to right. Of these time records, *c* applies to records *a* and *b*, *e* to *d*, and *g* to *f*. In *c* and *e* the larger oscillations divide the record into thousandths of seconds, or milliseconds, and the smaller oscillations divide each millisecond into sevenths. In *g* the time is marked by the 60 cycle current. May I add here that it will not always be possible to refer to time records. You will, however, be able to read them yourselves, if you will bear in mind that in some records the time is linear, in others, as in this one, logarithmic.

Record *a* shows the first three elevations, *alpha*, *beta*, and *gamma*, or the A group, of the action potential. The amplification for record *a* is such that the bar subtends 500 microvolts (μV). It should be added that, due largely to the unusually long distance of conduction, this *gamma*, and also certain other features to be mentioned, are more complicated than usual. Record *b* is the same as *a* except that the amplification has been trebled so as to bring *gamma* more clearly into view. At this amplification the height of *alpha* greatly exceeds the limits of the screen.

To demonstrate the next series of elevations it is necessary to get more time on to the screen, and to increase the amplification still further. For record *d* amplification is 20 times that used for record *a*, as may be seen by the bar, which here subtends 25 μV; and the transit speed of the spot of light, as indicated by the 1000 cycle oscillations of *e*, is now much slower than in records *a*, *b*, and *c*. As a result, *alpha* and *beta* are crowded into the first part of the record and are so high, and consequently so faint, that even such parts as remain within the limits of the screen are scarcely visible. The increased amplification is indicated by the increased size of *gamma*, the irregularities of which now give it a sawtoothed appearance. *Gamma* is succeeded by a declining potential, not an elevation, labeled *δ*. The decline here, though continuous, is not smooth. Some of the small deflections on it are constant and undoubtedly significant, while others seem to be inconstant. The decline, *δ*, terminates in another elevation, B. The end, however, is not yet. To show it (record *f*) the transit speed must be slowed still further, to the extent indicated by the 60 cycle time intervals of record *g*. It is then seen that, following the diphasic artifact of B, there still remains a slight and declining potential which runs into the final diphasic elevation, C. The declining potential following B again is roughened by irregularities some of which are too constant to be assignable to artifact.

A similar series of records from another bullfrog's sciatic nerve is shown in Fig. 5. These records are made by an older, the contact print, method and are neither as clear cut nor as clean as the first set shown. The dots here subtend milliseconds; their spacing falls off logarithmically as the record runs. The preparation differs from the first only in that the distance of conduction (9.1 cm.) is shorter, the lead being from a thicker part of the nerve. Under the circumstances the irregularities seen in the first set of pictures are glossed over here due to the larger number of fibers responding. *S* is the shock artifact and is to be disregarded. Records *a* and *b* are exactly alike except as regards amplification which for *b* was 18 times that for *a*. They show the A elevation with its *alpha*, *beta*, and *gamma* waves. Records *c* and *d* have the same amplification as *a* and *b*, respectively, but the transit speed of the recording spot is very much slower, as the time lines show. Elevation B is seen here with the declining featureless *delta* region between it and the preceding *gamma*. Record *e*, running out to 300 msec., i.e., 0.3 sec., exhibits the final, the C, elevation; elevations A and B are compressed to the left. A connected diagrammatic picture of this action potential, transferred to scale on to linear time coördinates from the logarithmic coördinates of the record, is seen in Fig. 6.

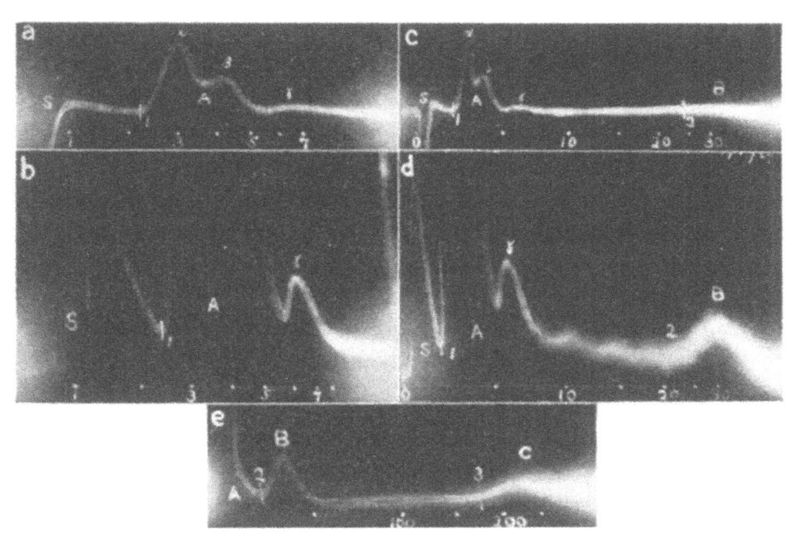

FIG. 5. Complete action potential of the bullfrog's sciatic nerve; distance of conduction 9.1 cm. *S* is the shock artifact. The starts of the A, B, and C elevations are indicated by *1*, *2*, and *3*, respectively. The dots give the time in msec., zero (sometimes ahead of the picture) being the start of the induction shock eliciting the action potential. For further description see text.

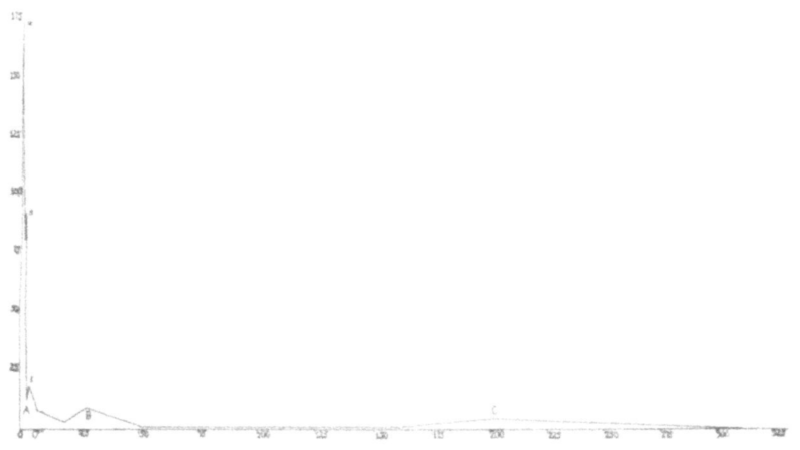

FIG. 6. The action potential of Fig. 5 transferred to linear coördinates. The starts, crests, and notches of the A, B, and C elevations, in their correct temporal positions, are joined by straight lines, the start of A resting on zero. Ordinates, amplitude in mm.; abscissae, time in msec.

The start of the compound action potential is placed on zero time. The base line gives the time in milliseconds and the ordinates relative voltage. To the left is seen the A elevation with its *alpha*, *beta*, and *gamma* waves and its *delta* region, the last not labeled. It is followed by B, its long tail, and a long-drawn-out C elevation. The durations apply, as will become clear, to the particular distance of conduction obtaining in this case.

Factors Determining the Configuration of the Conducted Action Potential

The action potential we have described is the one that results when, as has been said, the initiating shock is made strong enough to stimulate all of the fibers of the nerve. The evolution of the picture, as the strength of the stimulus used to evoke it is gradually increased to that point, can now be traced; and to be concrete, we shall give the readings made while recording the first of the action potentials shown, that of Fig. 4. If a value of 1 be assigned the strength of the shock which evoked a just perceptible *alpha* elevation, then as the strength was increased the height and the area of the *alpha* elevation increased very rapidly and became complete, i.e., *alpha* became maximal, when the strength reached 1.5. From 1.5 to 1.6 there was relatively little increase in the total area of the action potential. Then from 1.6 to 2.9 *beta* formed and attained its maximum, and from 3.3 to 4.45, *gamma*. While the strength was being increased from 4.45 to 11.7 the successive increments of potential diminished more or less steadily so as to form the *delta* region; but at 11.7 the increments of potential with increments in stimulus strength began to increase again, until the strength reached 53.3; it was this increase that formed B. Following B, there were the three minor increments forming at undetermined strengths, and then nothing definite developed until the final increment began that formed C; this started at a strength of about

100 and C was maximum at about 300. It should be borne in mind in this connection that the range of the values obtained in such a procedure depends in part on the configuration and the duration of the stimulating current. Indeed, there are those, e.g., Lapicque (1908) and Wyss (1932), who maintain that even the sign of the range can be changed by adequate electrical stimulation. In any event, it is obvious that, as tested, the fibers of the sciatic have a wide range of excitability.

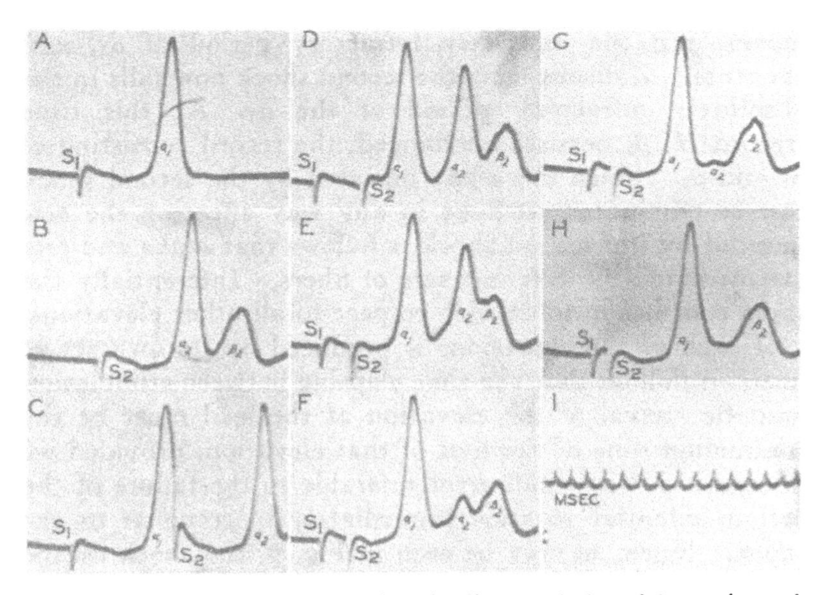

FIG. 7. Records demonstrating that the fibers producing *alpha* can be made refractory without affecting the fibers producing *beta*. The time (*I*) is linear.

When the characteristic elevations of the conducted action potential in the frog's sciatic were first encountered it took some experimenting to convince ourselves of the correctness of our original surmise that each is formed by contributions from a different set of fibers. A number of tricks were played on the nerve in order to prove that such actually is the case. Fig. 7 illustrates one of them, the only one it will be necessary to relate here. Two shocks are sent into the nerve at different times through the pair of electrodes placed on its central end. The earlier of the two shocks (S_1 of record *A*) is given a strength that is just maximal for ***alpha*** (it produces α_1) and the later shock (S_2 of record *B*) a strength that is just maximal for ***beta*** so that α_2 and β_2 record. S_1 and S_2 are then delivered in succession, and in successive trials S_2 is moved earlier as in records *C* to *H*. As the two responses approach each other the second ***alpha*** (α_2) becomes smaller and smaller as it moves into the relatively refractory period of α_1, and eventually α_2 disappears: the second shock now falls in the absolutely refractory period of the α_1. At this time (record *H*) β_2 persists unchanged, the record consisting of α_1 and β_2. Since the ***alpha*** initiated by the second shock can be eliminated without in any way affecting the ***beta*** induced by the second shock it follows that ***alpha*** and ***beta*** are mediated by different sets of fibers. Inferentially the same conclusion holds with respect to all other elevations.

If each of the elevations is produced by its own set of fibers, it follows that the time elapsing between stimulation and the arrival of the elevation at the lead must be the conduction time of the foot of that elevation, provided we disregard a very small error referable to the failure of the action potential to start immediately in response to the shock. Since, as may be seen in Fig. 7, the shock escape as well as the action potential can be made to appear on the record, conduction time is readily measured. So, when the distance of conduction is known, it becomes a matter of simple arithmetic to calculate the rate of conduction of each of the elevations. A set of typical values, obtained again from the nerve supplying the records of Fig. 4, is as follows in meters per second at room temperature—***alpha***, 41; ***beta***, 22; ***gamma***, 15; B, 4; and C, 0.7.

The calculation of conduction rate assumes that the foot of an elevation is propagated along the length of the nerve at a uniform rate. In the parts of the nerve we have used in the experiments thus far described this certainly is justifiable. That it is, can be demonstrated in a number of ways. One of the methods we have used is illustrated by Fig. 8. A number of pairs of stimulating electrodes are arranged along the nerve at appropriate and known distances from the lead, and records are made of the action potential resulting from stimulation of the nerve through each pair in turn. The figure shows the result obtained when the stimulus employed is just maximal for *beta*. With the widest separation of stimulating electrode and lead, giving a conduction distance of 143 mm., the two elevations (α and β of the lowermost record) are distinctly separated from each other. As the distance of conduction is shortened to 84, 41, and 21 mm. in the successive trials, giving the records mounted above each other in this order, the elevations become more and more completely merged, and, as they overlap, their potentials sum. In the uppermost record, where the conduction distance amounts to

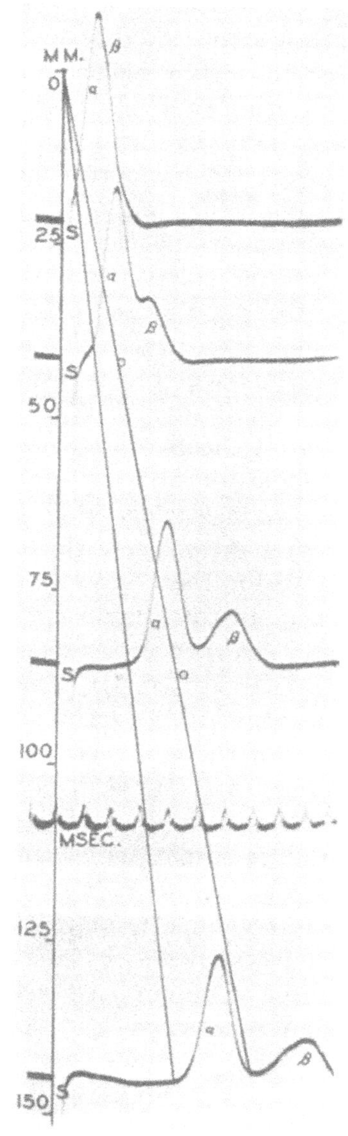

Fig. 8. Demonstration of linearity of the velocity of propagation of *alpha* and *beta*.

21 mm., *beta* is discernible only as a slight hump on the down stroke of the action potential near its crest. These records have been mounted over each other with the initiating shock escapes, *S*, on the vertical line, and spaced vertically by distances which are proportional to the respective distances of conduction. A straight line drawn through the start of the lowermost action potential and the point marking zero distance of conduction, exactly intersects the starts of all of the other records.

The start of *beta* is obvious only in the lowermost record where it is practically out in the clear. Another line has been drawn connecting the start of this *beta* with the zero of conduction distance. The positions of the starts of *beta* of the two intermediate records have been located by calculation and are indicated by the circles. In this way it becomes clear that the second straight line intersects reasonably closely the starts of all of the *betas*. Since the time for these records is linear, this exhibit signifies that through this distance of 143 mm. the conduction of *alpha* and *beta* proceeds linearly. Some investigators (Marshall and Gerard, 1933) have described a decline in conduction rate toward the periphery. This and other records signify that if there is such a decline it does not occur within the range of the present preparation.

It can be seen in these records that as *alpha* and *beta*, and this is equally true of all other elevations, proceed along the nerve they broaden and become lower. The same change is more obviously apparent in Fig. 9. This figure shows a set of records, this time of *alpha* only, at three different distances of conduction, increasing from above downwards. Planimeter measurements of this set of records shows that despite the change in configuration the area of the elevation remains constant within the limit of the error of measurement. It may therefore be concluded that each of the elevations is formed of the sum of a large number of smaller elevations, such as might come of

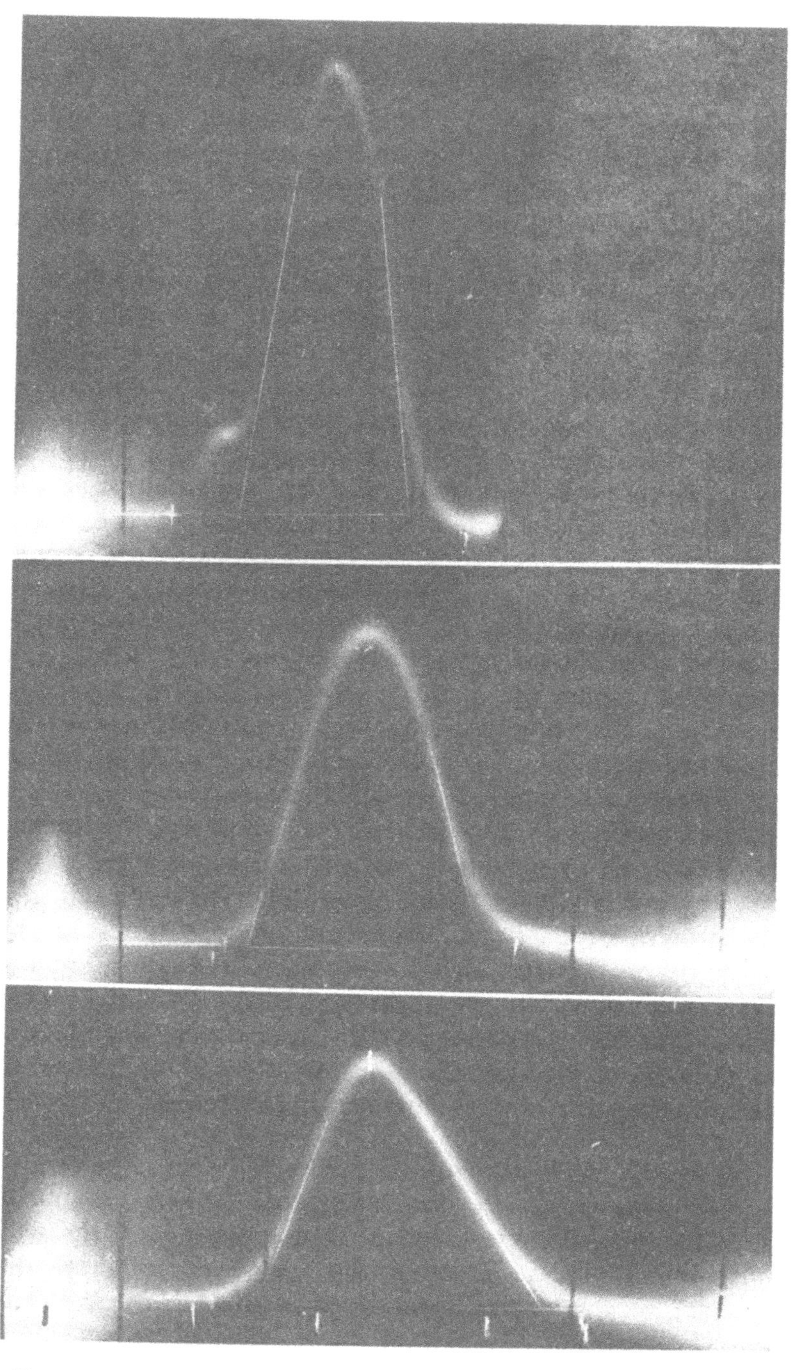

FIG. 9. *Alpha* action potential wave in the sciatic of the bullfrog recorded 14.5 (top), 44.0 (middle) and 85.0 (bottom) mm. from the stimulus. *x* is the shock artifact. The time indicated by the marks on the base line is in msec.

the potentials of individual axons, traveling at a range of rates that would produce the spread seen here. The idea, put into graphic form, is seen in Fig. 10. Five axon spikes, each represented by a triangle, start simultaneously to run from scratch (zero on the vertical scale), but each runs at a different speed so that they scatter as they go. Being potentials, the picture they produce at any instant of the race is their sum. The diagram shows that this (the upper curve in each position) becomes a lower and a broader figure as the race proceeds along the vertical scale. Without going further into detail it may be concluded that the picture of the conducted action potential of the frog's sciatic nerve signifies that the latter contains fibers conducting at rates ranging, probably continuously, between, say, 42 m.p.s. and 0.3 m.p.s. In this fiber "spectrum,"[2] however, the fibers apparently are not distributed uniformly in respect to conduction rate. The appearance of secondary elevations on the action potential cannot be produced merely by the presence of fibers of widely different velocities of conduction but there must be variations in the numbers of the fibers at each velocity. A similar conclusion was reached, it will be recalled, relative to the distribution of fibers with respect to their excitabilities: the growth of potential apparently is continuous as the strength of the stimulus is increased, though there are exacerbations at the strengths which bring out the elevations. A large mixed nerve then, such as the sciatic, contains fibers of every conduction rate throughout the range and of every excitability between the extremes.

Conduction Rate, Spike Form, and Fiber Size

Now the sciatic nerve is composed of fibers ranging in size between about 20 and 1 *microns* (thousandths of a millimeter). Is it possible that both excitability and velocity of conduction are related in some manner to the

[2] The term was first used by Heinbecker, Bishop, and O'Leary (1934).

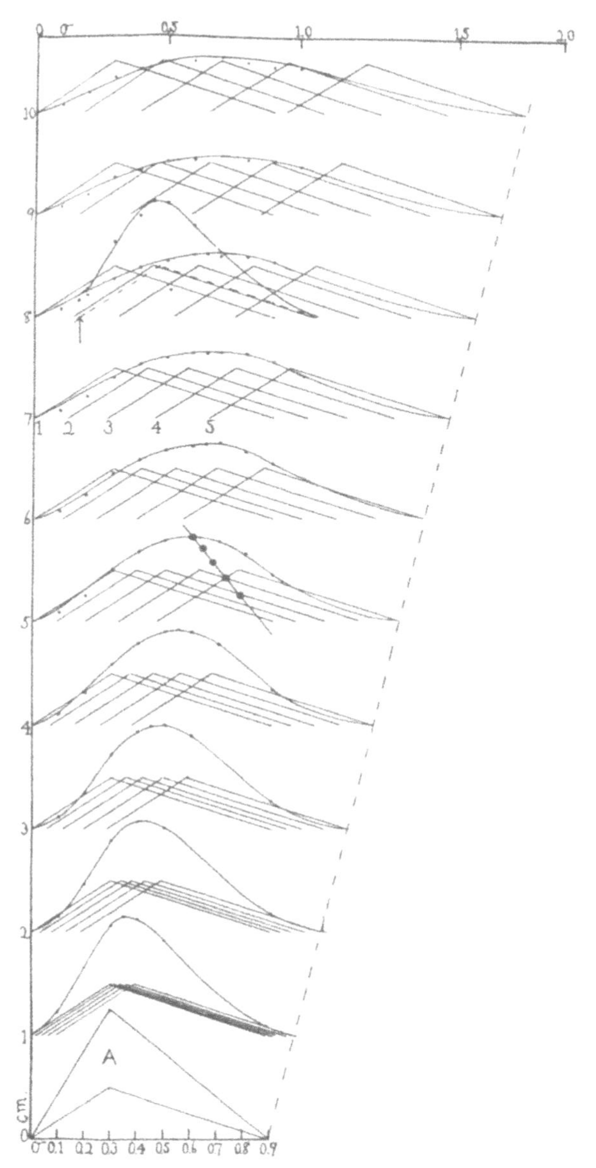

Fig. 10. Diagram illustrating the nature of the change in the configuration of an elevation on the assumption that the latter is made of the sum of axon spikes traveling at different velocities.

sizes of the fibers? The idea that there might be a relation between fiber size and conduction rate arose long before it was realized to what extent the conduction rates in fibers vary. The suggestion was first made by Göthlin in 1907 when he developed his cable hypothesis of conduction in nerve fibers, now regarded as untenable. The Lapicque school, also, came to the conclusion that there is a relation between fiber size and conduction rate. This conclusion, though, was based on evidence which they interpreted to mean that the chronaxie of a fiber is a function of its size (Lapicque and Legendre, 1913) and also of its conduction rate; but, as will be explained later, improved methods of investigation show that there is no simple relation between chronaxie and conduction rate.

In 1927, at a time when, of the compound action potential, the only known elevation was A, with its three definite waves, *alpha*, *beta*, and *gamma* and a dubious *delta*, and when it had not yet become possible to record with the electron oscillograph the responses of single axons, the attempt was made to ascertain by calculation and experiment whether there is any relation between the physical dimensions of fibers and their conduction rates (Gasser and Erlanger, 1927). Let me repeat for emphasis that at that time only *alpha*, *beta*, and *gamma* were known; and they were then believed to account for the contributions of all of the constituent fibers to the action potential of the nerve. In other words, we then were trying to fit the potentials of all of the fibers into only a small part of the action potential they produce.

In approaching the problem it was argued that if fiber size and conduction rate are related it should be possible to reconstruct graphically the conducted compound action potential of a nerve as of a given distance of conduction, if certain values were known. The values include, of course, (1) the diameters of each of the responding fibers and (2) their conduction rates; in addition (3) the amplitudes and

(4) the temporal configurations of their action potentials or spikes. The means whereby this information was acquired at that time, and the method of employing it in the attempt to solve the problem, can best be conveyed by considering in some detail a specific case, and we might as well continue to confine our attention to the peroneal extension of the bullfrog's sciatic.

The outside diameters of *all* of the 854 medullated fibers constituting a nerve were measured microscopically at the locus on the nerve from which a record had previously been taken of an action potential conducted a known distance. The measured diameters were recorded in the manner seen in Fig. 11. Each dot in the lower graph represents a fiber measured. The values along the horizontal axis of the graph indicate the diameters of the fibers in *microns*. The number of dots in a vertical row therefore is the number of fibers of the indicated diameter. For example, the four dots on the vertical over 15 signify, in effect, that there were in the preparation four fibers measuring 15 *microns* in diameter.

The spikes of these axons were all assigned the same intrinsic amplitudes and the same durations. In so far as concerns amplitude, the basis for the decision was entirely arbitrary. However, equality of intrinsic potential amplitude does not mean that the extrinsic effect produced by the potential on the recording mechanism is the same for all fibers; on the contrary, experimental evidence indicates that the recorded potentials will vary as the cross-sectional areas of the fibers. That relation, therefore, was adopted. In order to minimize the amount of necessary arithmetic all of the fibers falling within a range of 0.5 *micron* were regarded as having the same diameter. Then the amplitude of the "spike" produced by such a reduced fiber was taken to be proportional to the square of the diameter and to the actual number of fibers composing it, or to D^2n.

The D^2n values thus determined are indicated by the treads on the staircases of the lower graph.

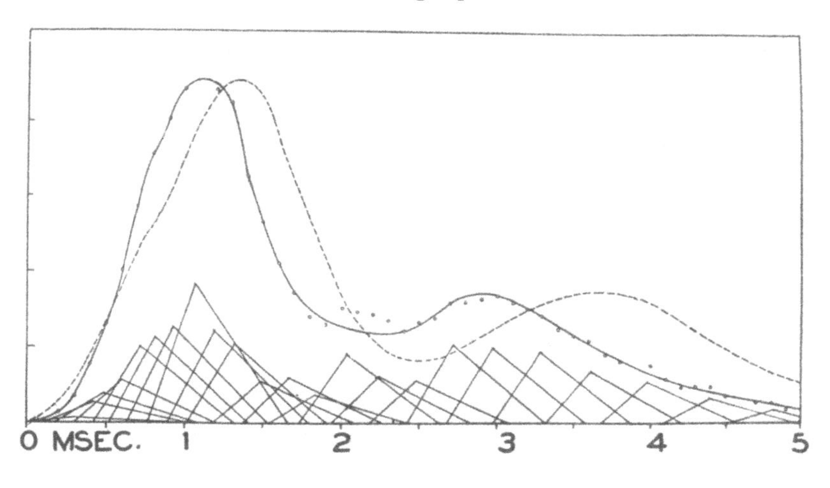

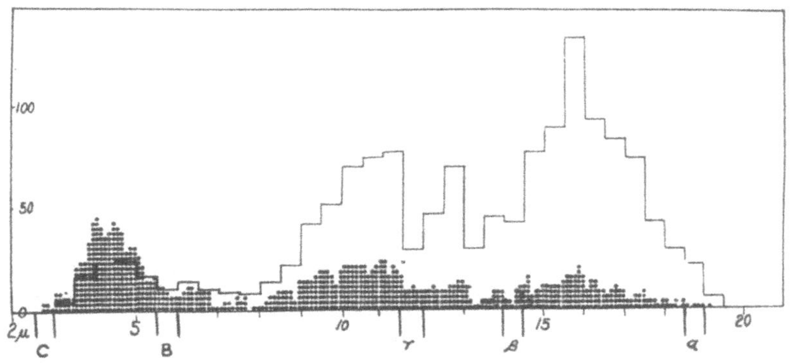

FIG. 11. Lower graph: Fiber-size map at the locus of the lead from a bullfrog's peroneal nerve.

Upper graph: The continuous curve, indicated by circles, is the action potential reconstructed on the basis of the assumption that conduction rate varies as fiber diameter; the broken curve shows the ***alpha*** and ***beta*** elevations as recorded.

Coming now to temporal configuration, at the stage of progress then obtaining it was believed, firstly, that all axon spikes (only A potentials were known) had essentially the same time constants and that therefore, secondly, a lead from, or close to, the point on a nerve at which the action potential is initiated by a shock would give the time constants of all of the spikes. But, as we shall see in a moment, modes of attack that have since become available disclose the fact that axon spikes are not all alike. Moreover, it can be shown that what is needed for these reconstructions is not the configuration of the axon spike at the site of its initiation, but rather the configuration of the spike that is conducted to the lead.

It is perhaps unnecessary to indicate why in the earlier experiments devised to determine the duration of the axon spike (Erlanger and Gasser, 1924; Bishop, 1927), the lead was from the stimulated locus; for it is obvious that if any conduction were involved, differences in the conduction times of the responding fibers would result in the arrival of their spikes at the lead over a period corresponding with the lag of the slowest propagated disturbance behind the fastest, and that the recorded figure, therefore, would be correspondingly broader than the axon spike proper. It should be added, moreover, that the record taken at the stimulated locus actually is not the record of the unconducted action potential, since the action potential conducted away from the stimulated point leaves behind a trail of potential which protracts the recording of the active locus (Erlanger and Blair, 1934). Be this as it may, the time values selected in making the original reconstructions, namely, 0.3 msec. for the time to maximum of the axon spike and 0.6 msec. for the decline from maximum, are not far from those now accepted as applying to the conducted spikes of the fastest of the A fibers; but they apply to those alone.

Finally, as has been seen, the conduction time of the fastest fiber contributing to the compound action potential can be readily determined; it is the conduction time of the action potential itself. If, then, the assumption be made that the fastest fiber is the largest fiber and that the conduction rates in the other fibers vary as some function of their sizes, all four of the factors will be available that are needed for the reconstruction of the compound action potential. After having tried several functions one was found which yielded reconstructed action potentials which seemed to match fairly satisfactorily the recorded action potentials. The relation that seemed to work was that the conduction rate varies as the diameters of the fibers.

A reconstruction based on these principles is seen in the upper graph of Fig. 11. Each triangle represents one of the synthetized D^2n spikes described above. The "spike" produced by the fiber measuring 19 *microns* is the one that would arrive at the lead first since its conduction time is the shortest. All other "spikes" would lag behind the fastest each by an interval determined by the difference between its calculated conduction time and the determined conduction time of the fastest spike. Therefore the triangle representing the 19 *micron* fiber is set with its start on zero and the other triangular "spikes" later each by its own conduction lag. And each "spike" (or triangle) is given an amplitude that is proportional to the height of its tread in the lower graph. Granting all of the premises, the sum of these triangles should be a figure similar to that of the record. The synthetic result is shown by the dots through which the *smoothed* solid curve is drawn; and the dotted curve is the *alpha* and *beta* part of the conducted action potential as recorded, transferred to the graph to scale. The two curves are reasonably alike, and it therefore was concluded that "*to a first approximation* the velocity of a fiber is determined by the diameter." In this particular case the crests of the elevations on the record are later than the apparently corresponding crests on the *smoothed* reconstruction, and the difference is greater the slower the elevation. Such, however, was not always the case. Moreover, a satisfactory *gamma* wave never did emerge from the reconstructions made in this manner.

There is, however, a more serious difficulty to be met. It was stated above that to form *alpha, beta,* and *gamma* by this procedure it seemed necessary to requisition all of the medullated fibers of the nerve. So when higher amplification disclosed the presence of B and C elevations, at least one of which had to be referred to medullated fibers (since it is not likely that both could be formed by nonmedullated fibers), it became obvious that the initial assumptions required revision. One way out of the difficulty would be to assign different properties to each of the groups of fibers, to A fibers, to B fibers and to C fibers, while holding to the initial fiber-size rule within each of the classes (Erlanger and Gasser, 1930; Gasser, 1935*a*). Another way would be to test the initial assumptions, and, if they were found wanting, to search for justifiable factors which could serve as a basis for another attempt to ascertain whether there actually is a unitary relation between fiber diameter and conduction rate.

The clue to another fiber-size rule, applicable apparently to the entire fiber spectrum, was found during the course of experiments on single axons in frogs' nerves, and it now becomes necessary to describe some of those observations in order to prepare the way for the further development of this theme. There is scarcely any need of telling this audience that the first successful technique for limiting the potential record of a nerve to that produced by a single one of its fibers was developed by Dr. Bronk in collaboration with Adrian (1928). Through dissection of a stretch along the course of a nerve they destroyed by trial the continuity of all but one of its fibers between the site of origin of the action potential and the locus of its registration.

In our study of single axons we have made use of preparations and of a technique which permit one to record and to compare the individual responses of a number of axons of a nerve. For the most part the preparation employed has consisted of the sciatic nerve of the green frog with the ramification attached that extends to the tip of the third digit. This we have designated the phalangeal preparation. The fine end of the preparation is but 20 to 50 *microns* in diameter and contains only 20 to 100 fibers with diameters ranging from 12 *microns* down to those of the smallest nonmedullated fibers. The distal end of the nerve is so fine that moistening the nerve chamber by usual methods does not suffice to protect the preparation from drying. It is necessary, in addition, to keep moistening the nerve itself. To accomplish this the finer part of the preparation is mounted vertically in the moist chamber. In this position it crosses the slit-like orifices of the mercury-calomel electrodes to which it is held by the force of capillarity. Drops of Ringer's solution, flowing into the chamber from without, fall along the vertical nerve, frequently while the nerve chamber is open and infrequently during observations.

The ratio of the cross-section of the largest fiber to the cross-section of the phalangeal nerve is so large that an unusually large fraction of a fiber's potential can be recorded. Single fibers have been encountered which have yielded potentials higher than 300 microvolts. With the amplification available (up to 2,000,000) the deflection produced by such a fiber can be given a height of more than 5 cm. Recorded spike voltages run from that maximum down to the noise level of the amplifier. From the peripheral end of a satisfactory preparation, when the nerve is stimulated at its central end, from 6 to 8 conducted axon spikes can be recorded which are high enough and sufficiently separable to be available for present purposes. It thus becomes possible to compare the properties of a number of axons under strictly comparable conditions. However, the axon spikes cannot be assigned to the anatomical units, to the particular fibers, producing them. Granting, though, a relation between fiber size and conduction rate, and it can scarcely be doubted that some relation obtains, one can, by determining the propagation rates of the several spikes, correlate the latter with fiber size in relative terms.

The axon spikes obtained from such a preparation are seen in the records of Fig. 12. The numeral under the rising limb of each gives the rising, or crest time in milliseconds, and the numeral to the left of the rising limb, the

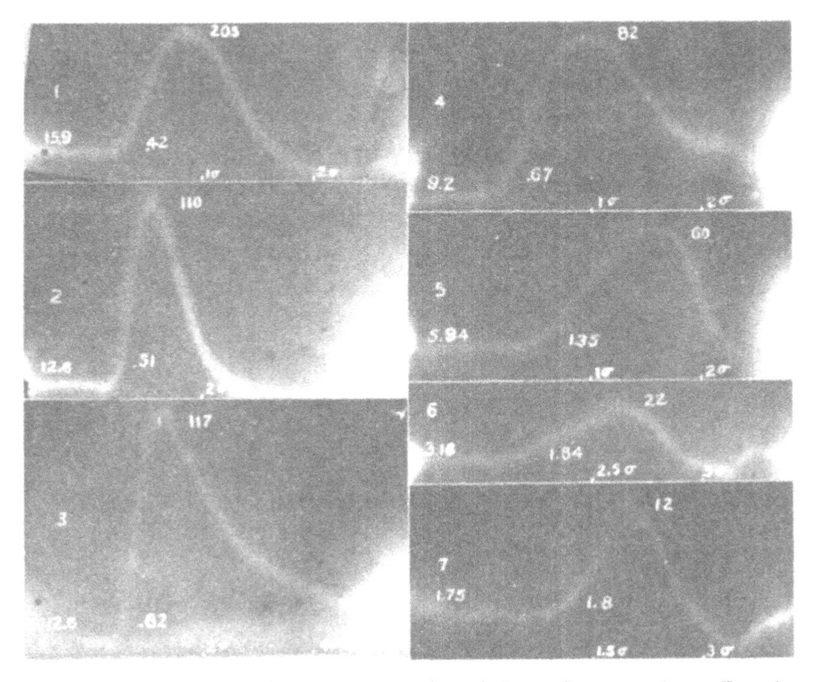

Fig. 12. The spikes of seven axons of a phalangeal preparation. For the significance of the numerals see text.

conduction rate in meters per second. The amplitude in microvolts is given by the numerals opposite the crests; the latter show how the amplification was increased as the spike voltage decreased. It will be noted that the amplification needed to record the slowest of these spikes is approaching the point where amplifier noise becomes high enough to be disturbing. For this reason it has not been possible to obtain accurate data for spikes traveling much under 2 m.p.s. The usual steps were taken to make the responses monophasic. Nevertheless, the slowest spikes all have been diphasic and so, too, have some of the faster spikes. In the latter case, however, the diphasicity may not cut down time to maximum or amplitude, whereas the slower spikes are seriously modified thereby.

The data from this and from another experiment like it are collected in Table I where it is seen, to take one case,

TABLE I

Velocities, Crest Times and $V \times CT$ of the Spikes of Axons in Two Phalangeal Preparations, Those of 12-31-33 Being Supplied by Fig. 12

Exp.	*Velocity*	*Crest Time*	*Product*
1-16-33	16.2	0.39	6.32
	15.3	0.43	6.58
	11.1	0.49	5.44
	11.1	0.53	5.88
	5.98	0.63	3.77 Diphasic
12-31-33	15.9	0.42	6.68
	12.6	0.51	6.43
	12.6	0.62	7.81
	9.2	0.67	6.16
	5.94	1.35	8.0
	3.16	1.84	5.81
	1.75	1.80	5.81 Diphasic

that of the nerve yielding the records of Fig. 12, that the time to maximum increases from 0.42 msec. to 1.8 msec. as the conduction rate slows from 15.9 to 1.75 m.p.s. The Table supplies also the explanation of this increase in time to maximum. The last column lists the product of the crest time and the rate of conduction; examination of those values shows, if we exclude the spikes which obviously are cut down by diphasicity, that this product tends to be constant. This result signifies that the length of nerve subtended by the potential developed in a fiber is the same for all fibers; and there seems to be every reason for believing that this should be the case, at least when, by adjusting the amplification, the spikes all are brought to about the same height; the spread of potential from the active locus then appears to be the same for all. From the standpoint of the synthesis of action potentials, which is our present concern, this result tells us that the spike durations selected must increase as the conduction rate slows.

The pictures of the axon spikes (Fig. 12) show also that the voltage amplitude (numerals opposite crests) is lower the slower the fiber. The results from a number of the experiments are collected in Fig. 13 where it can be seen

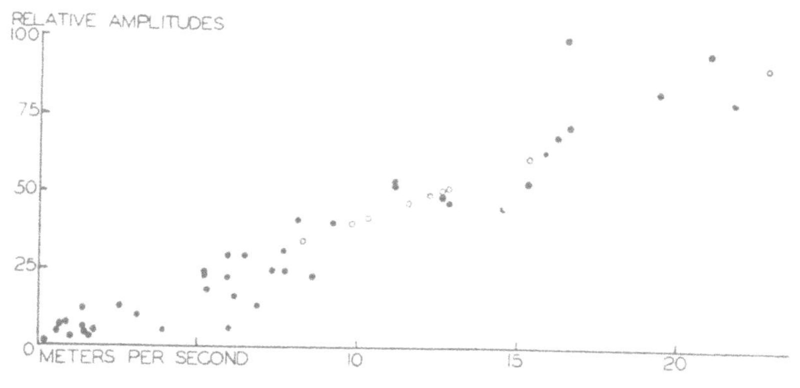

Fig. 13. Amplitude of axon spikes (ordinates) plotted against conduction velocity. Data from a number of preparations brought to comparable amplitudes.

that the spike height falls off linearly with the conduction rate. Since the slower fibers undoubtedly are also the smaller fibers this result is not entirely surprising, but by means of simple arithmetic it becomes possible to deduce from the fact a relation between conduction rate and fiber size which is of some interest of its own account and of prime significance in respect to the reconstruction of action potential pictures.

If, as we have assumed, the *intrinsic* potential of action attains the same amplitude in all fibers, and if the whole nerve acts in effect as a homogeneous shunting resistance, it follows that the *recorded* potential of action, A, varies as the cross areas of the fibers, or as the squares of their diameters, D (Gasser and Erlanger, 1927). But we have just seen that the amplitude, A, of the recorded axon spike varies also as the conduction rate, C. Accepting these relations at their face value, it follows that the conduction rate varies as the diameter squared. Expressed algebraically we have:

$$A = kD^2 \quad \text{and}$$
$$A = k'C; \quad \text{hence}$$
$$D^2 = k''C$$

Synthesis of the Compound Action Potential

Examination of the responses of single axons leads, then, to two conclusions which are inconsistent with the assumptions made in connection with the first attempts to determine the relation obtaining between fiber diameter and conduction rate. They are, (1st) that the durations of conducted axon spikes not only are not all alike, but actually vary as the conduction time, and (2nd) that the conduction rate in fibers varies not as their diameters, but rather as their diameters squared. With these new clues as a basis and with the complete action potential picture at our disposal, attempts have again been made to reconstruct action potentials from the original fiber-size analyses. The results obtained, we may say in advance, have not been wholly convincing; they are, however, interesting enough to report.

In order not to protract the discussion inordinately we are going to start with a bias in favor of the view that conduction rate varies as the square of the diameter. The mean conduction rates of *alpha*, *beta*, *gamma*, B and C of the sciatic nerve have been, in m.p.s., 42, 25, 17, 4.2 and 0.7. Now with what sizes of the fibers plotted in Fig. 11 do these rates correspond? The largest fiber of the nerve measures 19 *microns*. The only one there is of this size, however, could scarcely produce a visible potential. Let us say that only when the 18.5 *micron* group enters the picture does the foot of the action potential become perceptible. Then, on the assumption that conduction rate varies as the diameter squared, we find by calculation that the beginnings of *beta*, *gamma*, B and C correspond with fiber diameters, in *microns*, 14, 11.5, 5.5 and 2.4, respectively. The positions occupied by these *fibers* in the map are marked. Reading from right to left, each of the marks γ and B falls exactly at the start of a fiber pile. The β mark falls in an indefinite sort of hollow in the chart, but not far ahead of a point (13.2 *microns*) where there is a very definite increase in the number of fibers. The C mark corresponds with the smallest of the medullated fibers; of this we shall have more to say in a moment.

Now, to become more exacting, can one synthesize a reasonable conducted sciatic action potential out of the fiber-size map of Fig. 11 by employing the principles previously used by us (Gasser and Erlanger, 1927), but modified by taking conduction rates that are proportional to the square of the diameters and the spike times as increasing in proportion to the conduction times? An attempt is shown in Fig. 14.

The reconstruction is drawn here in three tiers for the same reason that three speeds were used in recording the whole of the sciatic action potential. To show the speedier parts the time units must be long, and the uppermost tier, it will be noted, covers only the first 7 milliseconds of the reconstruction. The second row is planned to show features of medium speed, and it covers the first 30 milliseconds. The lowermost tier covers the first 200 milliseconds, and even that does not suffice to bring in the C elevation; only its start shows.

The light triangles, as before, represent the axon spikes whose amplitudes are derived from the map of Fig. 11.

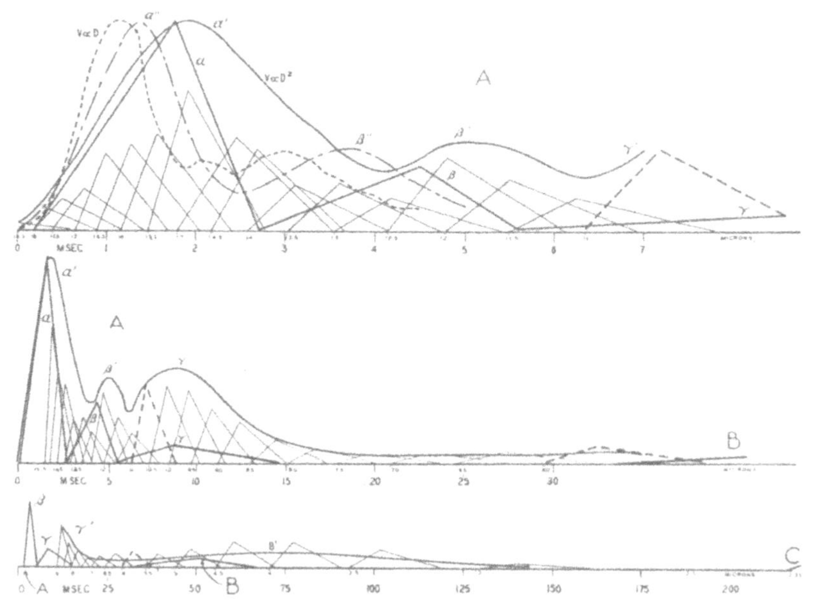

FIG. 14. Reconstruction of sciatic action potential (the curve indicated by α', β', etc.) based on (*a*) conduction rate varies as diameter squared, and (*b*) duration of spike potential varies as conduction time, employing data supplied by the lower graph of Fig. 11. The dotted curve labeled $V \propto D$ is, on a comparable scale, the reconstruction based on the linear relation of velocity to diameter; and the dot-dash curve, labeled α'', β'', etc., is the comparable action potential record (see Fig. 11, upper graph). The heavy triangles, labeled α, β, etc., are derived from Fig. 4 by joining points representing in time and height the starts and crests of the corresponding elevations. All data, where necessary, have been transposed so as to be as of a conduction distance of 9.1 cm., excepting the troughs between the triangles α, β, etc., which have been dropped to the base line.

Their sum forms the synthetic sciatic action potential, the solid line curve labeled $V \propto D^2$, α', β', and all of the other primes. The labeling serves to indicate that the synthesis brings out waves which resemble *alpha*, *beta*, *gamma*, and

B; it would probably have developed an elevation resembling C also, had statistics for the nonmedullated fibers been available.

For comparison with the synthetic curve, the record of the sciatic action potential of Fig. 4 has been drawn into this picture to scale (see legend) in the form of the heavy triangles labeled α, β, etc., after making an adjustment needed to bring the reconstruction and the record to comparable distances of conduction. That there are differences between the two is obvious enough; and yet when all of the difficulties are considered, some of which will be mentioned in a moment, the similarities are, we believe, sufficiently close to justify the premise on which the construction is based, namely, that the velocity of the impulse in a fiber varies as the square of the diameter. The most obvious deviation is a temporal one—the crests of the recorded elevations are earlier than those of the corresponding elevations of the synthesized curve (the prime curve). In addition, the troughs of the recorded action potential are relatively too deep, and consequently the descents from the crests are too steep. All of these are deviations such as would be produced by concealed diphasic artifacts, for which allowance cannot be made in reconstructions.

A word is needed by way of comment on the surmised relation to this scheme of nonmedullated fibers, which are not represented in our reconstructions. If the view of Duncan (1934) be accepted, to the effect that fibers whose diameters remain less than a certain value do not acquire medullary sheaths, it follows that the last pile of medullated fibers must be succeeded immediately by a nonmedullated pile. Duncan gives 1 to 2 *microns* as the critical diameter above which all nerve fibers are medullated and below which all are nonmedullated, but this in mammals. The smallest of the medullated fibers on the fiber-size map have a diameter of 2.9 *microns*. This, however, includes the medullary sheath. Since the sheath makes up about one-third of the total diameter (Donaldson and Hoke, 1905), the diameter of the unsheathed fiber corresponding with the 2.9 *micron* (the smallest) medullated fiber becomes about 1.9 *microns*, a figure which falls within Duncan's limits. Now the usual C velocity is 0.7 m.p.s. Proportionality of this velocity to the square of the diameter gives a diameter of 2.4 *microns* if the fiber were medullated, and this figure is of the order of magnitude, namely, 2.9, that is indicated by our fiber-size map. Upon the whole, therefore, the result of this synthesis seems to justify the conclusion that all fibers, nonmedullated as well as medullated, conduct at rates that are determined by the squares of their diameters.

It is necessary to admit, however, that we have other cases in which reconstructions on the new basis have presented deviations from the corresponding recorded potentials which have been wider even than those seen here. It has usually been in the *alpha*, *beta*, *gamma* range that the discrepancies have been most marked; moreover, it is usually in the A region of the fiber-size maps that it has been difficult to find features which could determine the elevations exhibited by the records. We shall have occasion to refer to such cases in the next lecture. On the basis of our experience, it would not be surprising if it should develop that there is in living nerve a sharper segregation of fibers relative to diameter than is displayed by our maps. However this may be, it is certain that all of the errors of measurement are in the direction of covering up discontinuities in fiber size. Nevertheless, failure to get a perfect fit should temper our conclusion; there may be factors in addition to diameter which modify the velocity of conduction in fibers. That there is room for difference of opinion is indicated by the fact that other investigators (Douglass, Davenport, Heinbecker, and Bishop, 1934), through similar fiber-size analyses and syntheses of a variety of nerves, have reached the conclusion that "conduction rate varies as a function of size between the linear and the square of the diameter." We shall, however, have occasion in the next lecture to consider additional evidence bearing on this question, most of which, we believe, supports the view that it is the cross section of the fiber and not its linear dimension that determines the velocity of propagation.

II

THE COMPARATIVE PHYSIOLOGICAL CHARACTERISTICS OF NERVE FIBERS

WHEN we watch the growth of the action potential of a large mixed nerve as the strength of the stimulus is gradually increased, as we did in the first lecture, the definite impression is gained that the preparation contains nerve fibers of every excitability in the range between the highest and the lowest; and of every conduction rate in the range between the fastest, of, say, 42 m.p.s. and the slowest, of less than 0.3 m.p.s. (in the frog).

In these respects, and thus observed, the properties of the fibers seem to vary in continuous series, like the wave lengths of light forming a complete normal spectrum. However, though the nerve spectrum appears to be without gaps, it obviously is not equally intense throughout the range. Due to an uneven numerical distribution of the fibers through the scale, the potential developed after conduction, when all of the fibers are activated simultaneously, has for each nerve a characteristic configuration. The conducted action potential of the sciatic of the frog, we have seen specifically, presents features, elevations, which are perfectly distinctive. In the part of the action potential first disclosed by the electron oscillograph three elevations were present invariably, and very rarely there seemed to be a fourth; these were labeled *alpha*, *beta*, *gamma*, and *delta*. Subsequently (Erlanger and Gasser, 1930), when higher amplification became available, and it was discovered that there was still more action potential to be labeled, the *alpha-beta-gamma-delta* region collectively was labeled A, and two additional, prominent later elevations, or groups of elevations, were labeled B and C. Moreover, the designation *delta* was then abandoned because, with the higher amplification, that particular region of the sciatic action potential turned out to be featureless, though not, it should be emphasized, devoid of potential.

The Question of Fiber Types

These tentative designations were applied not only to the elevations but also to the fibers which were believed to be the ones whose spikes combined to form the correspondingly designated features. Thus we spoke of an A elevation and of A fibers; of a B elevation and B fibers, and so on. Though the designations were regarded as tentative, it was, of course, believed that the elevations traveling at more or less comparable rates in different nerves were built of the spikes of fibers related to those in the sciatic from the standpoint of size. Moreover, while it was realized (Erlanger, 1927) that the slowest conducting fibers would include the nonmedullated fibers, there existed at the time no authority for concluding that the slowest moving of the elevations, namely, C, is formed by nonmedullated fibers only. It was possible to believe at that time that there might be large nonmedullated fibers conducting faster than small medullated fibers, and minute medullated fibers conducting more slowly than the ordinary nonmedullated. Therefore C was regarded merely as an elevation composed of the slowest conducting elements, the smallest fibers of whatever kind. It has only been since 1934 that one could with any assurance maintain that the C elevation in the sciatic of most species might be formed exclusively by nonmedullated fibers. For it was then that structures supposed to be nonmedullated fibers of large diameter were definitely shown to be conglomerations of nonmedullated fibers, and that whether a fiber is medullated or nonmedullated apparently depends only upon its size, fibers acquiring a medullary sheath when attaining a diameter measuring somewhere between 1 and 2 *microns* (Duncan, 1934).

By an approach from another angle and largely through a study of other than somatic nerves, other investigators (Bishop and Heinbecker, 1930) in the meanwhile had become convinced that the *delta* region is significant and characteristic, and they therefore employed another system of labeling, using A with its *alpha*, *beta*, and *gamma*, as before, but B_1 for the *delta* region, B_2 for B, and C, as before, for C. This classification assumed, moreover, that there are, corresponding with A, B_1, B_2, and C, four general types of fibers. "While the separation due to differing conduction rates," they said (Bishop and Heinbecker, 1930, p. 183), "is not always particularly sharp between any two given components of the potential, the interval of threshold between certain waves is more abrupt, and at these points there is still more abrupt and striking change in the duration of the absolutely refractory period, chronaxie and duration of the axon potential at the stimulated point." Each of these four fiber groups, they believed at first, was typified by a recognizable element of the nervous system, namely, the A group, by "large and medium sized, thickly myelinated fibers," the B_1 group, by "small and somewhat more thinly myelinated fibers," B_2, by "small thinly myelinated fibers," and C by "nonmyelinated fibers." [1]

Subsequently, however, these authors modified their views in certain respects. Thus they have stated (Bishop, Heinbecker, and O'Leary, 1933, p. 649) that "no discontinuity exists in the A and B_1 properties." They also say (Bishop, 1933, p. 467) that "all of the fibers of the optic nerve have relatively thin sheaths," and since they describe in the action potential of the optic nerve elevations which they regard as homologous with the A, B (which B is not stated), and C elevations of other nerves, it must be inferred that thickness of medullary sheath is no longer regarded as

[1] See also Heinbecker (1929).

a significant criterion for distinguishing fiber types. We are confirmed in this inference by the statement that "autonomic myelinated fibers and the smaller afferent depressor (cerebrospinal) fibers are indistinguishable anatomically . . ." (O'Leary, Heinbecker, and Bishop, 1934, p. 285). Finally, their conclusion (Douglass, Davenport, Heinbecker, and Bishop, 1934, p. 172) that "conduction rate varies as a function of size between the linear and the square of the diameter," implying, as it does, that all fibers obey this one law, can only mean that in their opinion fibers differ solely by virtue of diameter differences.

In the meanwhile the study of the properties of *individual axons*, referred to in the first lecture (Blair and Erlanger, 1933), had led to the conclusion that there is nowhere in the gamut of the sciatic fibers a demonstrable break in the continuity of the properties of fibers conducting at different rates (down to 1 m.p.s.) that can be regarded as significant. In the present lecture we propose to bring together and to discuss some of the evidence for and against fiber types in the peripheral nervous system. It is not our intention, in devoting a lecture to this subject, to present it in a spirit of controversy. Rather we are using the discussion of the question as a means of gathering together in a connected fashion a number of interesting observations which otherwise might appear to be unrelated. The discussion of this subject in the literature reminds one somewhat of a sentence in Claude Bernard's *Introduction to the Study of Experimental Medicine.* This master experimenter says there, "When two physiologists . . . quarrel, each to maintain his own ideas or theories, in the midst of their contradictory arguments, only one thing is absolutely certain: that both theories are insufficient, and neither of them corresponds to the truth."

The preparation, the phalangeal nerve, employed in the study of the responses and of the properties of individual nerve fibers (Blair and Erlanger, 1933) was described in the first lecture. It is necessary to add in this connection only that despite the minuteness of the nerve, presumably every functional type of fiber contained in the sciatic is represented in it, with the possible exception of voluntary motor fibers. We can be sure that it contains fibers belonging in all of the principal categories, A to C, mentioned above. This, of course, is a matter of some importance where a comparison is to be made of fibers ranging through the whole width of the spectrum. Let me recall, also, that the method employed in studying single fibers does not permit one to identify the properties observed with the particular anatomical units, the fibers, which they characterize. Instead, properties must be correlated with the fibers' conduction rates; conduction rate in other words is used as the base line for the comparison. But, as a matter of fact, conduction rate has some advantages over other possible common denominators. For example: (*a*) once an action potential is initiated by the induction shock, conduction is continued by a purely physiological mechanism; (*b*) conduction is a relatively stable process in that it remains constant over considerable periods of time; and (*c*) conduction rate can be easily and accurately determined by means of the electron oscillograph.

Branching Axons

In employing conduction rate as the base line one must guard against error that can result from unrecognized branching of fibers. Anatomical studies (Dunn, 1909) would lead one to anticipate branching in about 30 per cent of the fibers of our preparations. We have occasionally encountered evidences of branching in our records, and may digress for a moment to describe an instance of it, with a view to indicating how disconcerting branching could be if unrecognized. A record from a branching fiber is seen in Fig. 15. It shows three axon spikes (*a*, *b*, and *c*), all differing in height, and arriving at the lead at different intervals after the delivery of a single shock, the second spike about 1.5 msec., and the third about 3 msec. later than the first. One would naturally be inclined to take this picture to mean that the shock was stimulating three different fibers conducting at as many different rates. The evidence, however, is conclusive that the shock was stimulating but one fiber. Thus it elicits all three spikes, that is to say, the entire picture, or no spikes at all; and irrespective of the kind of stimulus employed, whether a brief shock or a rheobasic constant current, the spacing of the three spikes remains invariable. The spikes are not repetitive responses of an axon, for, as will become clear in another lecture, this picture bears no resemblance whatever to that of repetition. Unquestionably it is the result of the stimulation of the parent trunk of an axon which soon divides into three branches conducting at different rates. This observation, it may be added, is of more than passing interest, since it demonstrates that the rate with which a fiber conducts depends not upon type, but upon its size; for here we have three very different rates of conduction in offshoots from one parent trunk. How different the rates are cannot be definitely stated. A rough estimate based on differences in the amplitudes of the spikes (see Fig. 13) indicates that the ratio is as 4 is to 1; that is to say, if the fastest spike were traveling at the rate of 20 m.p.s., the velocity of the slowest would be about 5 m.p.s.

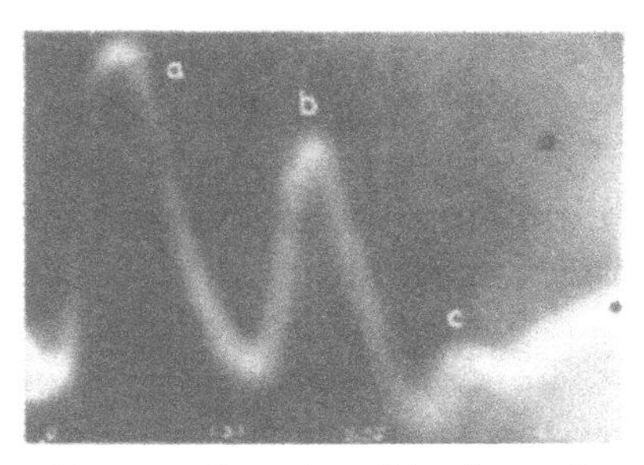

FIG. 15. Record resulting from stimulation of an axon that divides into three branches producing the spikes *a*, *b*, and *c*.

Using this instance of branching as an example, it is easy to indicate the confusion that branching could introduce into a statistical comparison of axon properties, when conduction rate is the common denominator. To make the situation as bad as possible let us suppose that the two most rapidly conducting of the three axon branches of Fig. 15, instead of continuing to the lead, as they do here, had turned off into nerve branches central to the lead. Then the record would have given no evidence whatever of branching, and there would have been no way of reconciling the combination of high excitability with slow conduction and low spike amplitude. Such departures from the rules to be developed, have in our experience been infrequent, less frequent rather than histological data would lead one to expect. Obviously, however, departures cannot be brushed aside excepting where it is possible to show that they actually are attributable to branching. In order to convincingly offset this complication, a very much larger accumulation of statistical data is needed than we have as yet made available; it still is barely possible that deviations have some significance. While referring to limitations of the method it should also be stated that only rarely have we been able to extend observations to fibers conducting more slowly than 1 m.p.s. This is due to the fact that the action potentials of slow fibers are so low that they cannot be accurately followed. With this discussion of the method in mind we are ready to proceed to a consideration of some of the data derived through a comparison of the characteristics of individual axons with their respective conduction rates.

Fiber Characteristics

The excitability range. In consonance with the results obtained from multifiber responses of large nerves it is found by this method that the excitability of fibers as tested by induction shocks and condenser discharges decreases as conduction rate falls off. In the case of the nerve from which the record of Fig. 16 was obtained, the spike that was conducted at the rate of 31.6 m.p.s. appeared at the lead when the relative strength of the induction shock rose to 9.2. At a strength of 23.8 the second spike appeared; it traveled at the rate of 16.5 m.p.s. A strength of 49.7 was required to elicit the remainder of the spikes, the fastest of which traveled at the rate of 11.1 m.p.s. Now it is not often that the spikes of a preparation are as regularly and as cleanly spaced as is the case here; much more commonly the scattering is irregu'ar, with partial or, occasionally, complete overlapping of spikes, so that often we have had to resort to devices of one kind or another in order to bring the individual spikes out into the clear. Graphs of the relation obtaining between conduction rate and threshold (as determined by induction shocks) of the available fibers in six different nerves are shown in Fig. 17. It is obvious that the points in each case tend to group themselves around a continuous curve of the hypobolic type.

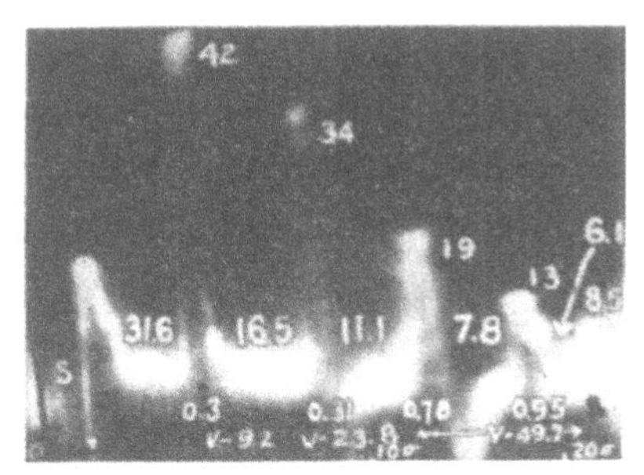

Fig. 16. Record showing the relation of fiber threshold (induction shocks, threshold strengths indicated by numerals following *V*) to conduction rate (numerals above base line of spikes).

If a rectangular constant current is used as the stimulus it becomes possible to ascertain for each of the available fibers of a preparation the strength of current needed to stimulate for every duration of current flow. In this way data can be secured which are suitable for plotting for each fiber the so-called strength-duration curve. Fig. 18 illustrates the result obtained from a nerve supplying six suitable fibers conducting at rates ranging from 25.7 to 2.01 m.p.s., as indicated. It is obvious that the more rapidly a fiber conducts, the lower is the voltage needed to stimulate it. There are no discontinuities on which could be based a division of fibers into types. Using values that can be read from such a family of curves one can plot the excitability for any duration of rectangular current against the conduction rate of the fibers of a preparation. Sample curves made in this manner are shown in the inset of Fig. 18. The curve labeled 0.02σ, the letter formerly used to designate milliseconds, expresses the relation obtaining between the voltage necessary to stimulate and the conduction rates of the different fibers when the rectangular

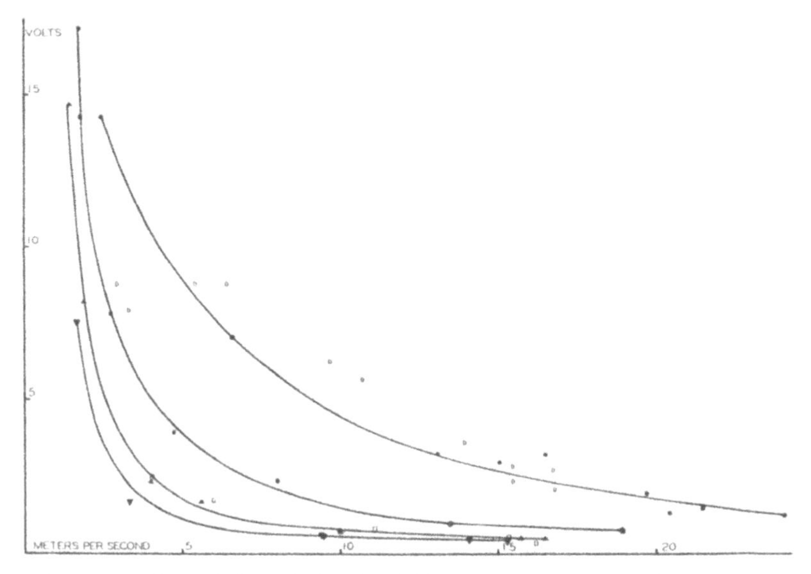

Fig. 17. The relation of the thresholds of axons as measured by induction shocks, to their conduction rates. Data from six preparations. The curve indicated by the *D* points has not been drawn. Ordinates, relative volts.

current has a duration of 0.02 msec. The other curves express the comparable relations when the current duration is 0.04 and 1.0 msec. and infinite, as labeled. Inspection of such a set of curves suffices to show that the relative excitabilities of fibers depends upon the duration of the exciting current. It can be seen, for example, that the range is much wider when the duration of the current is 1 msec. than when it is infinite.

Though the number of points on the individual curves is scarcely sufficient to designate accurately their configurations, it nevertheless is perfectly obvious, all things considered, that by this method, likewise, a relationship is

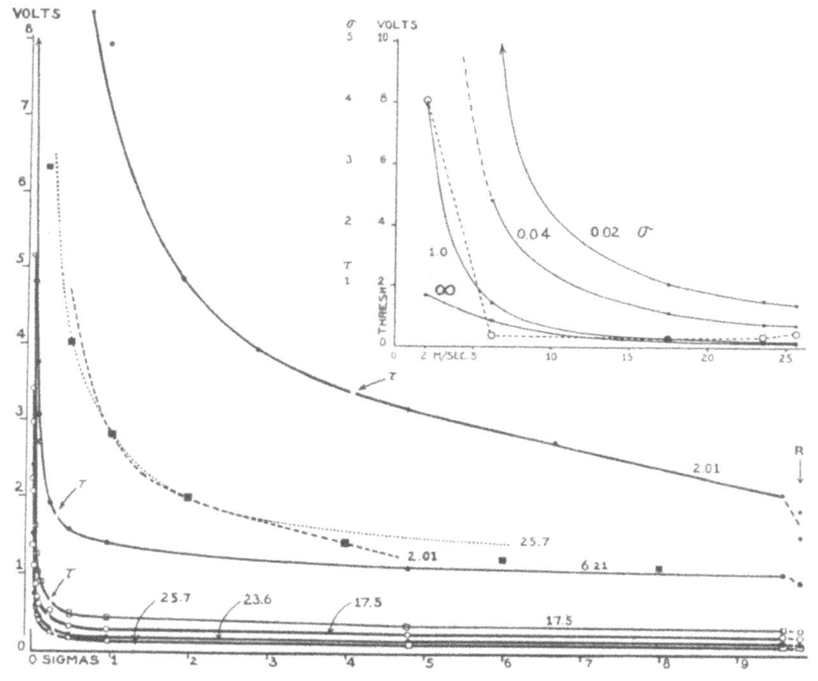

FIG. 18. Strength-duration curves from individual fibers of a nerve. Chronaxies (τ) are marked by breaks in the curves. The squares are selected points from the canonical curve of Lapicque; the dotted and the broken curves are the strength-duration curves of the fibers conducting 25.7 and 2.01 m.p.s., respectively, all plotted after comparable transpositions.

Inset. Continuous curves: Relation between threshold strength of rectangular currents of the durations indicated (namely, 0.02, 0.04, and 1.0 msec. and infinite) and conduction rate. Dotted curve: Relation between chronaxie and conduction rate.

demonstrated between conduction rate and fiber excitability that can be expressed by smooth curves.

Chronaxie. Any one familiar with the field that deals with strength-duration relationships of currents as stimuli would want to know how chronaxie varies with conduction rate. Chronaxie, it will be recalled, is a term that was introduced by Lapicque to designate the time a constant current must act if it is to stimulate when it is given twice rheobasic intensity—twice the strength of a threshold current of infinite duration. Chronaxie is believed by the Lapicque school to be a measure of the speed of excitability of tissues. Says Lapicque (1926, p. 356), the action potential travels 1 cm. per chronaxie.[2] If this were true, the chronaxies of fibers should increase as conduction rate decreases. As a matter of fact there is no such simple relation between chronaxie and conduction rate. The chronaxies of the six available fibers of this preparation are represented in the graph of the inset by the points connected by the dotted lines. The chronaxies of the fibers of this and of two additional nerves, plotted against conduction rate, are shown in Fig. 19. All of the curves combine to show that if there is any change in the value of chronaxie in fibers conducting at rates included between the fastest (here 26 m.p.s.) and about 10 m.p.s., it is in the direction of a slight diminution, whereas according to Lapicque it should increase with conduction time. Then, somewhere in the vicinity of 10 m.p.s. chronaxie begins to increase, slowly at first, but rapidly when the rate falls below 6 m.p.s.

These results differ, also, from those obtained in studies comparing the chronaxies of fibers of different "types," where, as in the case of Lapicque's observations, the responses were multifiber in character. Chronaxie, in the former case (Bishop and Heinbecker, 1930), was found to increase continuously as conduction rate slowed, but with exacerbations at the transitions from A to B_1 to B_2 and to C, and particularly from B_1 to B_2 (or at a conduction rate of about 5.5 m.p.s., in the bullfrog); whereas with single fiber responses the result is the one seen in Fig. 20. The

[2] Lapicque has since (1935) modified this conclusion and assumes that any given fiber has the same chronaxie throughout its ramifications.

latter result cannot, in our opinion, be taken to indicate that there are two types of fibers with characteristic chronaxies. But neither can one conclude from the result

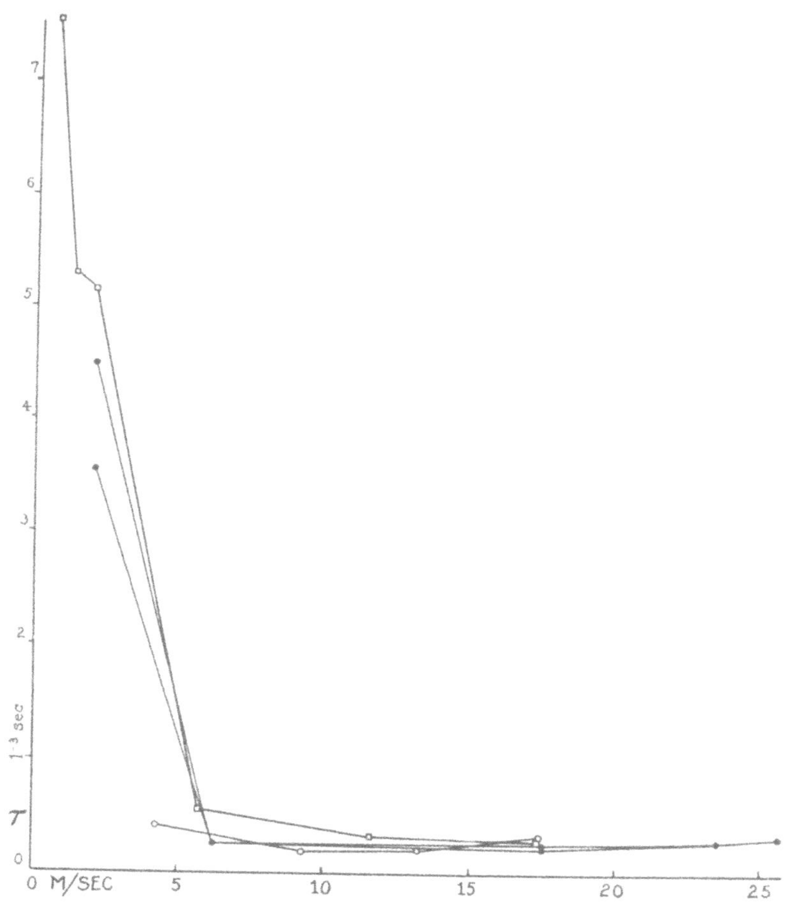

FIG. 19. The relation of chronaxie to conduction rate in the axons of three nerves.

that chronaxies disclose a simple uniform gradation of fibers with respect to their conduction rates. If there be a simple relation between chronaxie and velocity it is masked by conditions inherent in the method of measuring chronaxie.

Summation interval. If one stimulates a fiber with a shock that is below the threshold, the local excitability of the fiber remains above normal for a period that outlasts considerably the duration of the shock. This is the so-called summation interval. In another lecture we are going to have something to say regarding the nature of this

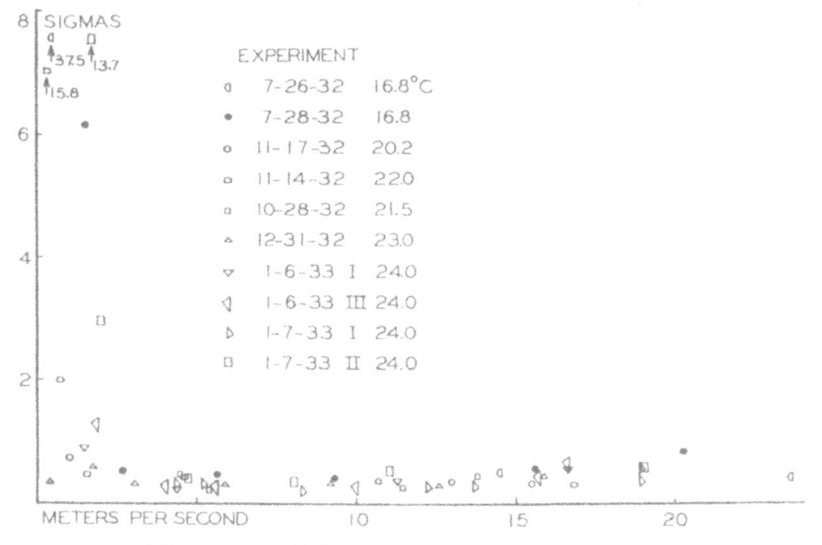

FIG. 20. The relation of the summation interval to conduction rate in the fibers of a number of preparations.

phenomenon and then will describe the method of determining it. To preface the present topic it is necessary to say merely that the summation interval has been supposed to be a measure of the speed of reaction of a tissue. This is a conclusion that rests on the observation made by Keith Lucas (1910) that tissues whose specific reactions, such for example as conduction in nerve, and contraction in voluntary muscle and in heart muscle, proceed at different speeds exhibit summation intervals which vary in duration in the same sense. Now, where the comparison is made between different tissues the determinations could be qualified by unavoidable differences in the conditions of the tests, such, for example, as differences in gross tissue resistance, in contact, and so forth. But where one carries out the determinations on the fibers of a nerve, the reacting units all are exposed to exactly the same conditions. If, therefore, summation intervals vary as the speed of physiological reactions, one might expect the summation intervals of the fibers of a nerve to increase as conduction rate slows. The results obtained from the fibers of eight different nerves are shown in Fig. 20, in which the data derived from each of the preparations are distinctively marked. It is seen there that in the conduction rate range extending down to about 2 m.p.s. the summation interval actually shortens. Below the rate of 2 m.p.s., however, it increases very rapidly in duration. It is very difficult, however, to appraise this result, since artifact probably enters into it very largely. The shortening of the summation interval may be the result of earlier onset of the "depression interval," which will be described in the next lecture; and subsequent lengthening perhaps is attributable to damage inflicted by strong shocks. At any rate, if it is maintained that these determinations disclose two types of fibers, it should be noted that the dividing conduction rate is in the vicinity of 2 m.p.s., or in the lower range of the fibers which contribute to B_2, and not in the range between B_1 and B_2.

Refractory period. The refractory period is another phenomenon which has been employed in the characterization of fibers. As is well known, a fiber, after responding to a stimulus, remains momentarily unirritable, or absolutely refractory, and then returns to normal through a relatively refractory period. In the routine method of measuring the period of absolute refractoriness the fiber first is stimulated with a shock (the conditioning shock) slightly above the threshold and then a second time with a shock (the testing shock) five times the threshold in strength and, while repeating the process, the interval between the two shocks is reduced until the second just fails to evoke a response. This interval is the absolutely refractory period.

But the results obtained by this method of comparing axons, like the one employing summation intervals, often are vitiated by the deleterious effects produced by the strong shocks that must be used in *testing* the less excitable and slower of the axons. A frequent consequence is that fibers conducting at rates slower than 6 to 7 m.p.s., yield absolutely refractory periods which are as long as, or even longer than, their relatively refractory periods. Obviously this is wholly anomalous, but it is unavoidable.

By the method described above, determinations have been made of the absolutely refractory periods of sixty-six axons in eighteen preparations at temperatures ranging between 19.3° and 26.3° C. All of the determinations are charted in Fig. 21 in relation to the rate of conduction. Each horizontal represents a nerve and the verticals erected on it the refractory periods of the fibers that were available for study. The length of each vertical is proportional to the duration of the refractory period, and the time scale for all is shown in the lower left-hand corner. The heavy verticals represent the absolutely refractory periods, the light verticals the relatively refractory periods. The chart makes it clearly evident that there is a tendency for the refractory period to increase as conduction rate slows. Owing, however, to individual variation among axons, it is difficult to appraise the rate of increase. One can state only in general terms that the usual increase in a preparation amounts to something over 60 per cent as the conduction rate decreases from the fastest down to 8 m.p.s. The increase in refractory time in general proceeds regularly, but more rapidly in the slower than in the faster ranges of

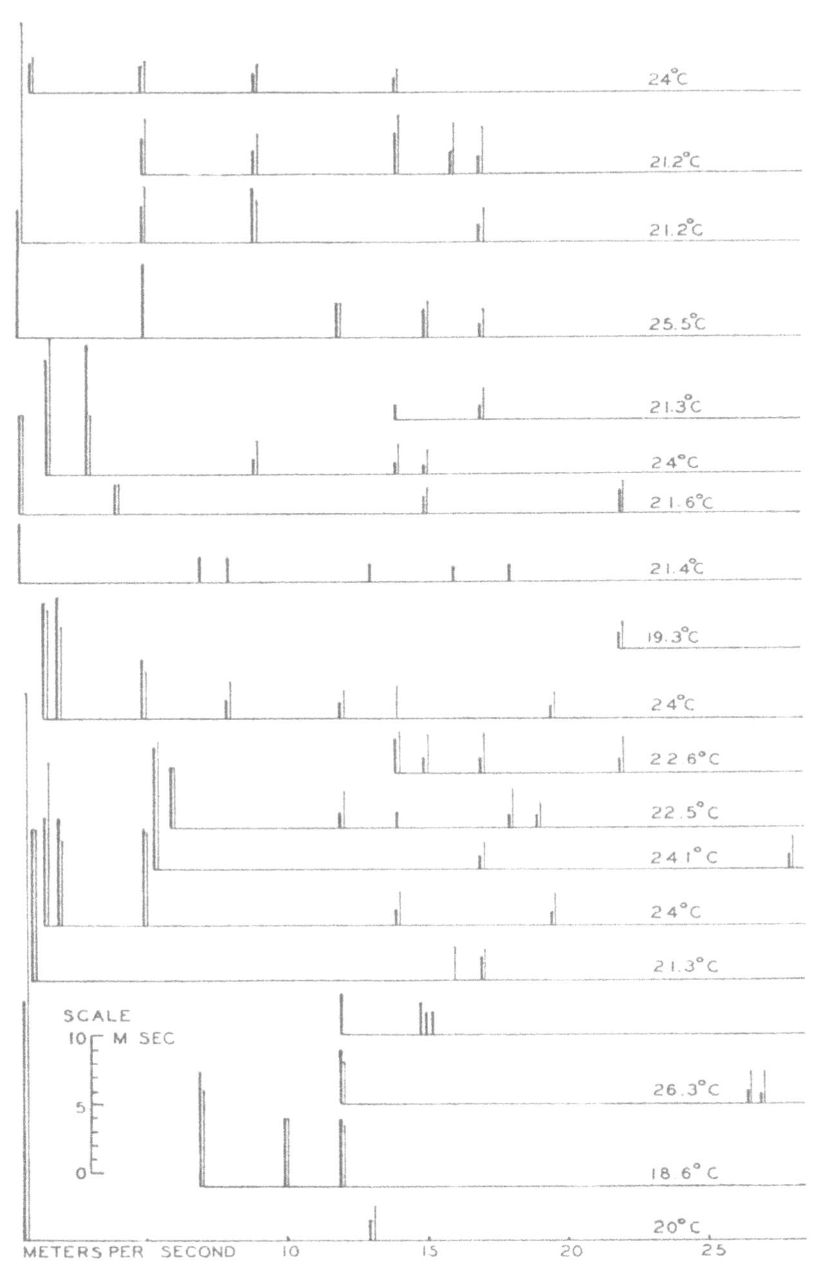

FIG. 21. The relation of the duration of the refractory periods, absolute (heavy verticals) and relative (light verticals), to conduction rate.

conduction; there are no obvious and consistent discontinuities. However, for reasons given, it has not often been possible to carry the determinations satisfactorily into the range of fibers conducting at B_2 rates (less than 5.5 m.p.s.). In the few satisfactory cases the refractory period has increased progressively.

Judging by the changing attitude of investigators with respect to the duration of the absolutely refractory period as determined through multifiber responses, it may be inferred that there is something about this phenomenon that is not completely understood. In 1924 we (Erlanger and Gasser) found in the *alpha* to *gamma* range an increase in refractory time amounting to 214 per cent. In 1927 (Erlanger, Gasser, and Bishop) we gave as the increase in the same fiber range, zero in some preparations, up to as high as 214 per cent in others, but expressed the opinion that the refractory periods of *alpha*, *beta*, and *gamma* fibers might all be alike. Writing in 1930, Bishop and Heinbecker convey the impression that they regard the absolutely refractory period as alike for all fibers in the A range, but in 1935 (O'Leary, Heinbecker, and Bishop) state (p. 649) that they find no indications of a "wide range of values" for the A fibers of mammalian nerves and in this connection mention specifically "0.7 to 0.9 *sigma*," or a range of 30 per cent. The comparable range in the frog, using the very sharp criteria available when dealing with single, discrete axon spikes, in our hands is, as has been said, about 60 per cent. We are convinced that no one is yet in a position to put his finger on the basis of these variable results; possibly polarization by demarcation potentials is a factor. In any event observations on single axons fail to disclose any sharp discontinuities in the duration of the refractory period along the fiber spectrum when the values are plotted against conduction rate. In general the relatively refractory periods vary in the same sense as the absolute periods.

Data on the *duration and amplitude of the conducted spikes* of individual axons of different velocities were presented in the first lecture. Since the product of the time to maximum of the spike by its conduction rate appears to be a constant, and since it is admitted by all that in large nerves there are fibers of every conduction rate, it follows that the time to maximum of conducted spikes must vary in continuous series through the fiber spectrum. Types cannot be distinguished on the basis of duration of conducted spikes. Neither can types be distinguished on the basis of the amplitudes of spikes, since amplitude diminishes linearly with conduction rate throughout the range down to 1 m.p.s. (see Fig. 13).

No attempt has been made to determine the duration of the spike potentials of axons at the loci of their initiation, that is to say, at the site of stimulation. It has seemed unnecessary to obtain such data because, as will be explained in another connection (Lecture III), the record obtained at the combined stimulating electrode and lead consists not only of the spike developing there but also of potential spreading back to the lead as the active locus moves away from it along the nerve. This criticism applies to all attempts of whatever kind (Erlanger and Gasser, 1924; Bishop, 1927; Gasser, 1931) that have been made to determine the duration of local activity by leads placed at, or close to, the stimulated point.

Observations on single axons of the sciatic nerve conducting at rates throughout almost the entire range fail, then, to supply evidence of the existence of discontinuities in their properties. Let us next, in our effort to ascertain whether there are different types of fibers, examine separately at their several sources the fibers that enter into the make-up of mixed nerves. As the prototype of mixed nerves we shall employ again the sciatic of the bullfrog. It scarcely is necessary to recall that fibers gain entrance to the sciatic through three general pathways. Certain of the dorsal spinal roots provide the path for the sensory fibers, and of the ventral roots the path for the motor fibers, and the supply of sympathetic fibers is derived via gray rami communicantes. Some of the motor roots, it should be added, contribute fibers to the sympathetic chain; those fibers course for a short distance in the sciatic trunks before leaving them via the white rami communicantes. Each of these avenues, we shall proceed to show, has a characteristic conducted action potential. After describing the latter, the attempt will be made to correlate them with the fiber constitution of their respective paths.

Motor and Sensory Root Action Potentials

For the study of conducted action potentials of spinal roots we may use preparations consisting merely of the excised root, or, if long distances of conduction should be desired, a preparation may be made such as is shown in Fig. 22. It consists of the sciatic nerve with one pair of roots attached. The sciatic may be stimulated and leads taken from either the motor root or the sensory root; or, if desired, conduction may be forced in the other direction.

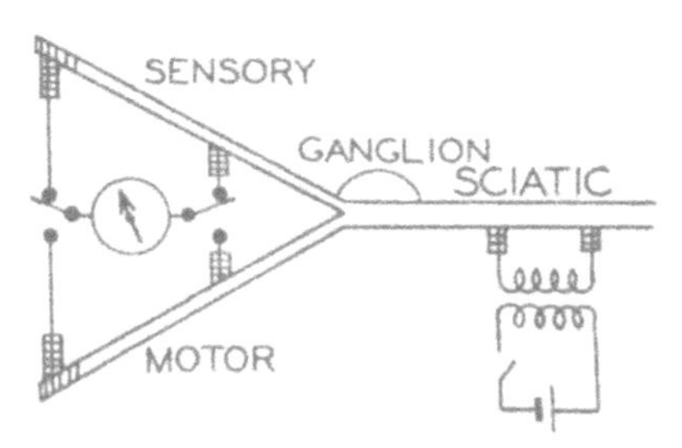

Fig. 22. Diagram showing the method of determining, under exactly comparable conditions, the motor and sensory contributions to the sciatic nerve.

The lower record of Fig. 23 shows the first elevation in a sensory root after long conduction. It matches in appearance the A elevation of the sciatic, the three waves on it resembling *alpha*, *beta*, and *gamma* in the sciatic action potential sufficiently to justify considering them as homologous, though there are certain minor differences between them.

To demonstrate the more slowly moving elements of the root action potential it is necessary, of course, to sweep the spot more slowly across the screen. In Fig. 24 are seen a number of sensory action potentials. Records *a* and *b* are from a sensory root (the 7th) with 6.5 mm. of conduc-

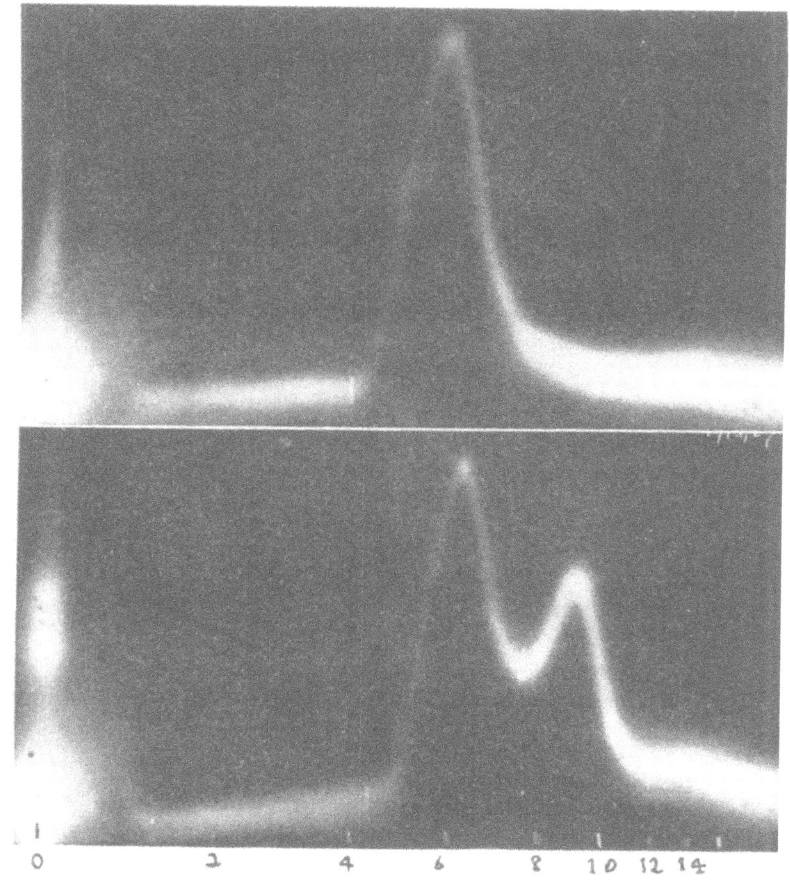

FIG. 23. The *A* elevation in a sensory (lower record) and corresponding motor (upper record) root elicited by stimulation of the sciatic; conduction distance long and the same for both. The time (msec.) applies to both.

tion; the only difference between them is that *b* is amplified 18 times more than *a*; the transit speeds of the two are so nearly alike that, practically speaking, comparable points on the two records are vertically over each other. In *a* the A elevation, because of slow transit speed, appears as a narrow spike, and at *C* there are faint indications of another elevation. In record *b* the higher amplification

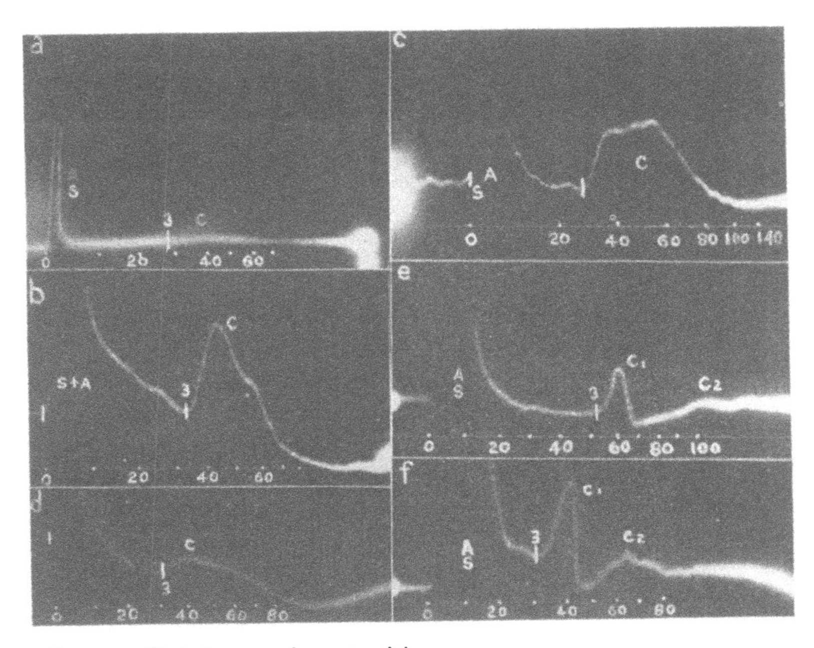

FIG. 24. Dorsal root action potentials.

a, 7th root; conduction distance 0.65 cm.; stimulus maximum for C; C rate 0.4 m.p.s.

b, same as *a* but with 18 times the amplification.

c, 9th root; conduction distance 1.2 cm.; C rate 0.47 m.p.s.

d, 9th root stimulated, lead from trunk central to rami; conduction distance 1.31 cm.; C rate 0.4 m.p.s.

e and *f*, 7th trunk stimulated, lead from dorsal root; in *e* the stimulus was applied distal to rami, in *f*, central to rami; rate, C_1, 0.74, C_2, 0.45 m.p.s.

discloses that C is a real elevation (with a conduction rate of 0.4 m.p.s.) and that the potential of the A elevation declines steadily until C rises out of it. It is, indeed, possible to show that the root contains fibers of every conduction rate, but we shall postpone the explanation of the method of proof until the same question arises for consideration in relation to the motor roots. The other records of Fig. 24 (see legend) show that both the A and the C elevations of sensory roots run into the sciatic nerve. If any of their fibers leave the nerve via the rami communicantes, they do not, in the roots we have submitted to examination, produce a perceptible elevation.

Now the question that interests us primarily is, what is the material basis of the configuration of the sensory root action potential? Fig. 25 is a size map of all of the medul-

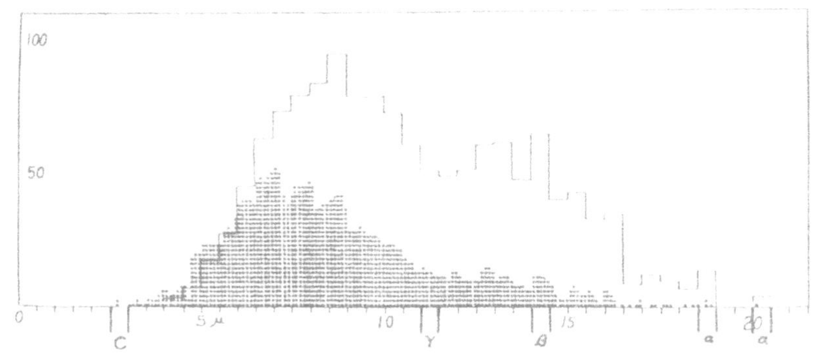

FIG. 25. Size map of all of the medullated fibers in a 9th sensory root of the bullfrog.

lated fibers (1577) of a 9th sensory root. We have no direct information relative to the nonmedullated content. According to Davenport and Ranson (1931), however, nonmedullated fibers exceed the medullated numerically over one and one-half times (this in mammals, however). And according to Duncan we might expect this mass of nonmedullated fibers to form a hillock on the map with its base in effect contiguous with the base of the last medullated pile. The largest nonmedullated fiber then would have an equivalent diameter of say 2.5 *microns*. These nonmedullated fibers could supply the basis for the C elevation. One cannot, however, discover so readily a basis for the *alpha*, *beta*, and *gamma* waves.

The first attempt, made in 1927 (Gasser and Erlanger), is reproduced in Fig. 26. This reconstruction was based on proportionality of conduction rate to fiber diameter and on temporal superimposability of all axon spikes. It yielded a curve as of 48 mm. of conduction (indicated by the dots) which bears some resemblance to the record (inset and undotted curve) obtained from the same nerve at a point

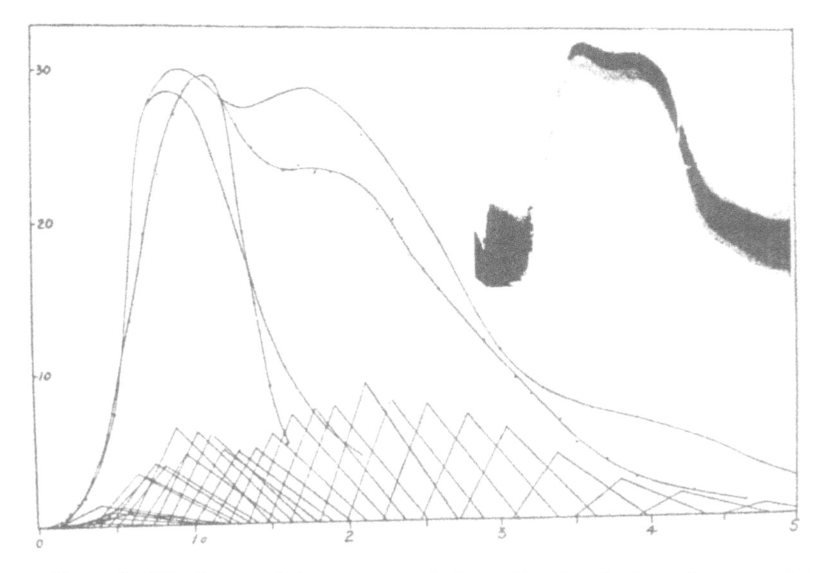

FIG. 26. The longer of the two curves indicated by dots is the action potential derived by reconstruction from the data of Fig. 25 on the basis of a linear relationship of diameter to conduction rate. The longer of the undotted curves is the comparable action potential derived from the record of the inset by converting logarithmic into linear time.

48 mm. removed from the stimulated locus; at least the curve exhibited elevations resembling *alpha* and *beta*, though, as usual, a *gamma* wave did not emerge. However, the reconstructions requisitioned for these waves all of the medullated fibers, and there are no fibers left, excepting nonmedullated, from which to derive the remainder of the action potential extending through to C.

Reconstructions of *sensory root* action potentials involve still another assumption, namely, that the peripheral and central branches from the ganglion cells have the same diameters, and there are reasons for believing that this may not be entirely justifiable, since Ide (1931) finds that the largest fiber tapers rapidly in the ganglion peripheralwards.[3] Then, there is to be taken into consideration the lag of impulses in traversing the ganglion (see below), though it is not likely that this alters significantly the configuration of the transmitted action potential. Under the circumstances, however, it does not seem worth while attempting a complete reconstruction of the sensory root action potential on the new basis.

A partial analysis, not particularly convincing in its results, can be made by proceeding somewhat arbitrarily, and selecting as the "fibers" forming the starts of *beta* and *gamma* those located where there seem to be appropriate changes in the contour of the fiber-size map. The "fibers" of 14 and 11 *microns*, labeled β and γ (Fig. 25), seem to meet this condition. The computations then depend, in addition, upon the selection of the fastest significant "fiber" of the root. The potential of the isolated 20 *micron* fiber would scarcely be perceptible in a multifiber record. The 18.5 *micron* "fiber" might be, and it therefore is taken to be the fastest "fiber." With these choices as a basis we have determined by calculation the propagation rates that would result if conduction rate varied *directly* as the diameter, and if it varied as the *square* of the diameter. The results are seen in columns 4 and 5 of Table II. For comparison with these values the usual conduction rates are included in the Table as column 3. Inspection of the Table shows that the values in column 5 are a far better match for the usual values than are those in column 4. We repeat, however, that the method of procedure leading

[3] Ide describes a similar tapering of the largest motor fiber.

Table II

Elevation	*Fiber diam.* μ	*Velocity m.p.s.*		
		Usual	*Computed*	
			$V \propto D$	$V \propto D^2$
Alpha	18.5	42	—	—
Beta	14.0	25	32	24
Gamma	11.0	17	24	15
C	2.5	0.4–0.5	6	0.8

to this result probably is entirely arbitrary; the result, therefore, cannot be regarded as conclusive.

The conducted action potentials of the sensory roots which we have investigated, the 6th to the 9th inclusive, all seem to have the same configuration basically. Motor roots, on the contrary, produce pictures which in one respect depend upon the root. But of that more in a moment. Let us start the description of the motor root action potential by referring to the figure (23) with which the discussion of the sensory root action potential was opened. The preparation, it will be recalled, consists of the sciatic nerve with one of its pairs of roots attached. The nerve is stimulated some distance from the lead, which for this picture has been shifted from the sensory root to an exactly comparable point on the motor root. The record (the upper one now) again shows only the A part of the response. In contrast to the exactly comparable sensory record seen below it, this action potential is simple in form, the simple down stroke terminating in a long-drawn-out tail. Comparison of the two records shows at once that sensory fibers participate, along with motor fibers, in the production of *alpha*, but alone determine the *beta* and *gamma* features of the sciatic action potential.

It can also be seen, to digress for a moment, that though these two action potentials start in the sciatic at the same instant, they reach their respective roots slightly asynchronously, the record from the sensory root beginning slightly later than that from the motor root. The difference is seen somewhat more clearly in Fig. 27, in which the upper record is from the motor root, the lower from the sensory; the stimulus is just maximal for *alpha*. This difference in the time of arrival in the roots is not assignable to inequality in conduction rates; actually it is due to a delay of the action potential of the sensory fibers in the sensory root ganglion. The lag, though small, amounting to only 0.14 msec. on the average, is nevertheless real; it develops, moreover, irrespective of the direction of conduction through the ganglion. Presumably it is referable to the manner in which the sensory fibers are joined to their ganglion cells at the T-shaped process, but just what it is that determines the delay is still the subject of speculation.

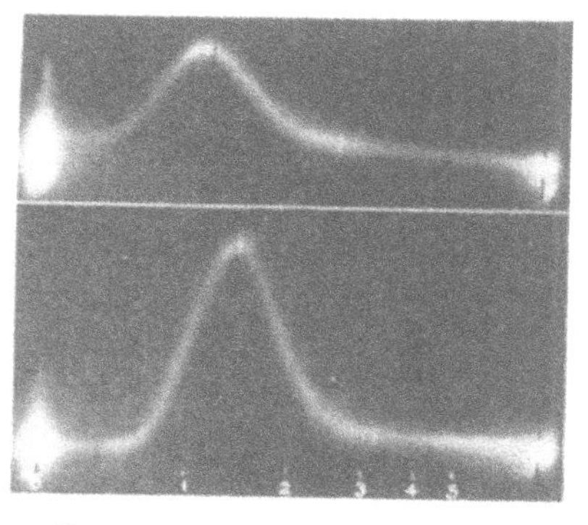

FIG. 27. The *alpha* elevation in the 8th sensory (lower record) and motor (upper record) roots elicited by stimulation of the sciatic; conduction distance 11.0 cm. The time applies to both.

When the preparation consists of the motor root alone, the conducted action potential of certain of the roots presents not only an A elevation, but also an elevation designated as C. In Fig. 28, *a*, *b*, and *c* all are records from the same root preparation (the 7th) with 11 mm. as the conduction distance. Record *a* shows the A elevation on a high speed line. By slowing the speed (record *b*) a C elevation (one traveling at the rate of 0.85 m.p.s.) is brought into the field. And by increasing the amplification and bringing C forward (record *c*), the configuration of this C elevation is made obvious. The rest of Fig. 28 is devoted to the demonstration of the fact that the C elevation of the motor root travels into the sciatic trunk but, with the minor exceptions, leaves the trunk via the rami communi-

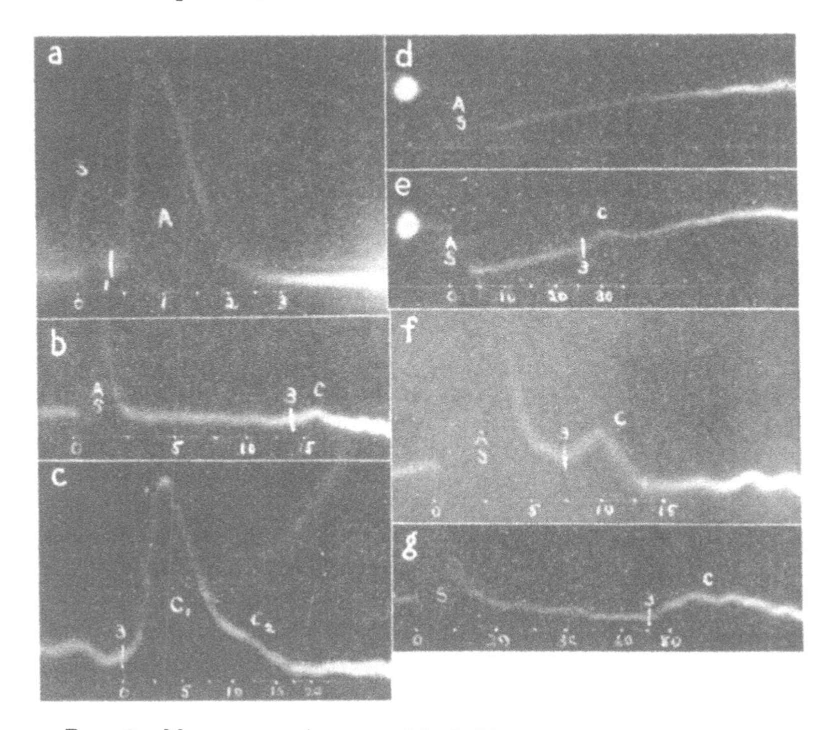

FIG. 28. Motor root action potentials, bullfrog.

a, *b*, and *c* are from the same, the 7th, root; conduction distance 1.1 cm. Amplification for *a* and *b* is the same, but the time, as indicated, is different. For *c* amplification is increased and elevation, *C*, is moved forward.

d, *e*, and *f* are from another preparation; the lead is from the root. In *d* the sciatic trunk is stimulated distal to the rami, in *e*, central to the rami; a C elevation is present in *e* only. In *f* the root itself is stimulated.

g. 7th root stimulated, lead from rami. In the same preparation stimulation of the dorsal root produced no deflection in the rami.

cantes instead of continuing on into the sciatic nerve. Since no demonstrable part of the A elevation takes this course, the fibers that make C must constitute the sympathetic outflow from the cord.

This conclusion is confirmed by the fact that for all practical purposes C appears only in those motor roots which are known to contribute to the sympathetic system.

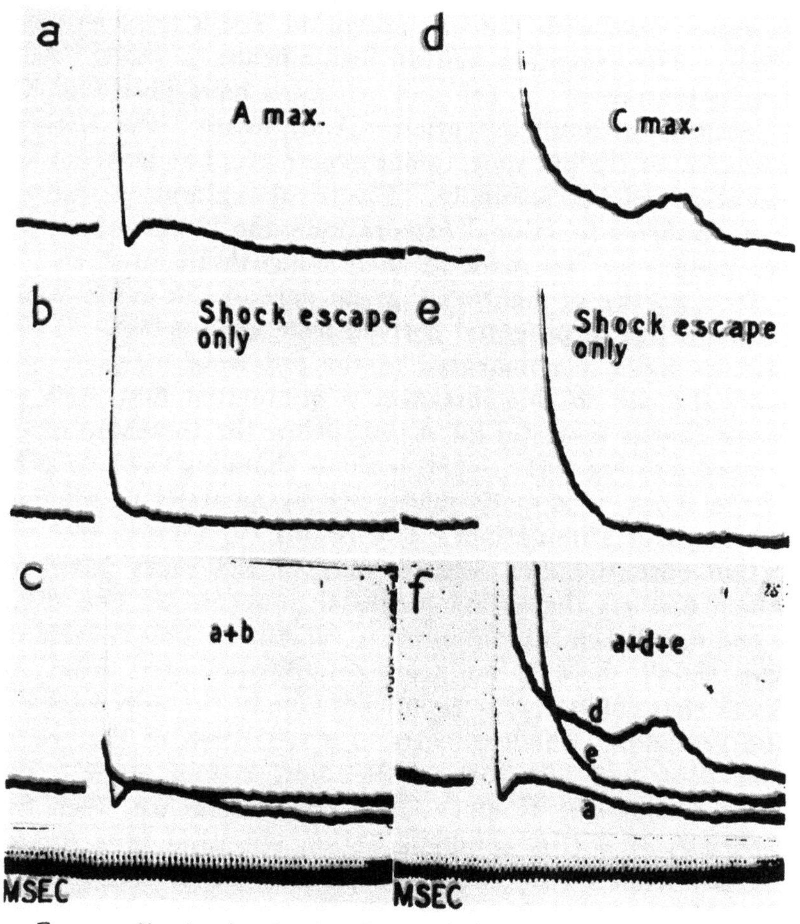

Fig. 29. Showing that there is *spike* potential in the motor root action potential throughout the entire conduction rate range. See text.

We have, however, explored but four roots, namely, the 6th, 7th, 8th, and 9th. It is known that peripheral stimulation of the 6th and 7th invariably elicits responses in tissues innervated by the sympathetic system, and of the 9th usually; whereas the 8th root never yields such a response. Now, fourteen preparations of the 8th root have been explored electrically. All have presented the usual fine A elevation and, with but one exception, not a trace of any other. The exception had an insignificant C wave. Action potentials of the 6th and 7th roots have presented C elevations without exception. And, finally, the action potential of the 9th root, in our experience, has lacked a C elevation only occasionally. Electrical exploration, therefore, confirms functional exploration—the C elevations of the motor roots are made by their sympathetic outflows.

It is worthy of mention that no part of the motor root record is free of potential derived from axon spikes. This fact we have demonstrated in the following manner. A root, the 9th in this instance, is stimulated first with a shock that is maximal for A, but below the threshold of C (record *a* of Fig. 29). Next, without changing the strength of the shock, it is made ineffective by anodally polarizing the locus of stimulation. The record (*b*) then is free of action potential and consists only of the shock artifact which distorts the action potential of record *a*. The two, *a* and *b*, are seen superimposed in record *c*. This combination shows there is no negative after-potential from A fibers that could play a significant rôle in the next part of the procedure, which consists of a repetition of the steps just described but with shocks now strong enough to elicit a maximal C elevation. The divergence, seen in record *f*, of *d* (the complete action potential plus shock escape) from *e* (the shock escape only), therefore, is assignable to spike potential. It thus appears that there is action potential in every part of record *d*. The whole fiber spectrum apparently is represented in the motor root action potential, and, as has been stated, the same method of procedure leads to the same conclusion relative to the sensory root also.

On the basis of this electrical analysis of motor roots one would expect them to contain medullated fibers of all sizes in a practically unbroken pile; aggregations of sizes, such as are believed to account for *beta* and *gamma* of the sensory root, should not be in evidence in fiber-size maps of motor roots. Fig. 30 shows that this is essentially true. However, in this particular case (a ninth root) fibers less than 7 *microns* in diameter were not measured. As a matter of fact there are in complete maps of all motor roots (7th, 8th, and 9th) medullated fibers extending down to 3 *microns* with some piling up of fibers around the 5 *micron* size. If there is any difference in these respects between the 7th and

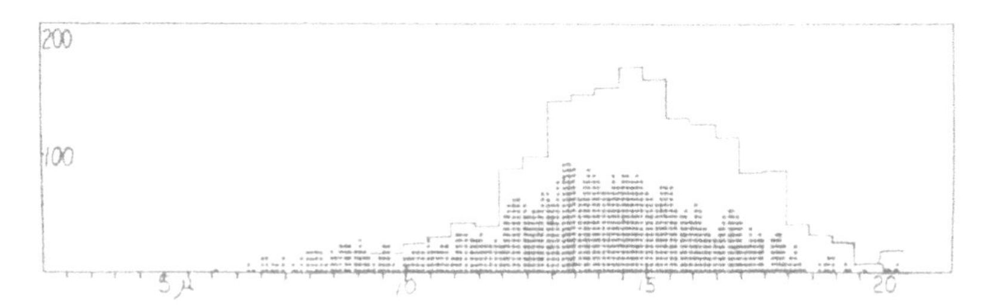

FIG. 30. Fiber-size map of the 9th motor root of the bullfrog. Fibers smaller than 7 *microns* not included.

8th roots it is insignificant. But in neither case do the smaller fibers seem to be sufficiently numerous to produce in the action potential an elevation as prominent as C. Moreover, on the D^2 basis such fibers could not possibly conduct more slowly than say 2.5 m.p.s. It seems necessary, therefore, to attribute the C elevation of motor root records to nonmedullated fibers, and yet it is positively stated (Davenport and Ranson, 1931) that motor roots (of mammals, however) contain relatively few fibers of that kind. Under the circumstances, one ventures to predict that, contrary to the view established by Gaskell in 1886 that fibers of the sympathetic outflow are of the small medullated type centering around a diameter of 4 *microns*, in the bullfrog the sympathetic outflow will be found to consist of nonmedullated fibers; if it does not we are dealing in the case of this C elevation with a wholly new phenomenon.

The small medullated fibers, we have stated, are almost if not quite as numerous in the 8th motor root as in the 7th. Since the 8th root has no sympathetic outflow the small fibers in that root, and therefore in the 7th also, must belong to the voluntary motor group. They probably add themselves to the group of small fibers which have been shown by Eccles and Sherrington (1930*a*) to make up a considerable fraction of the motor fibers in nerves going to muscles. These small fibers, according to Eccles and Sherrington, increase in number peripherally by fission. O'Leary, Heinbecker, and Bishop (1935) bring forward evidence, which, however, they do not themselves regard as conclusive, indicating that the small fibers are concerned with the innervation of the muscle cells of the muscle sense organs, the muscle spindles.

The Action Potential Contributed by the Gray Ramus

The sciatic nerve, it has been seen, receives from sensory roots fibers which contribute to the A and to the C elevations, and from motor roots fibers contributing to the A elevation, the sensory fibers covering the entire gamut and the motor fibers the gamut down to the part of the scale determined by nonmedullated fibers. The source of the fibers which produce the B elevation in the frog's sciatic still remains to be located. By exclusion, they must enter the nerve via the gray rami communicantes, since that is the only remaining source of supply; and electrical exploration has located them there. Fig. 31 will serve to indicate the method employed in determining the contribution of the gray rami to the action potential of the sciatic nerve. The preparation, diagrammed to the right, consists of one of the sciatic trunks, *T*, formed by the union of a pair of roots, and its rami communicantes, *R*. The fibers entering the nerve from the sympathetic system take the course indicated by the arrow. The lead into the oscillograph is from the peripheral end of the trunk, and ar-

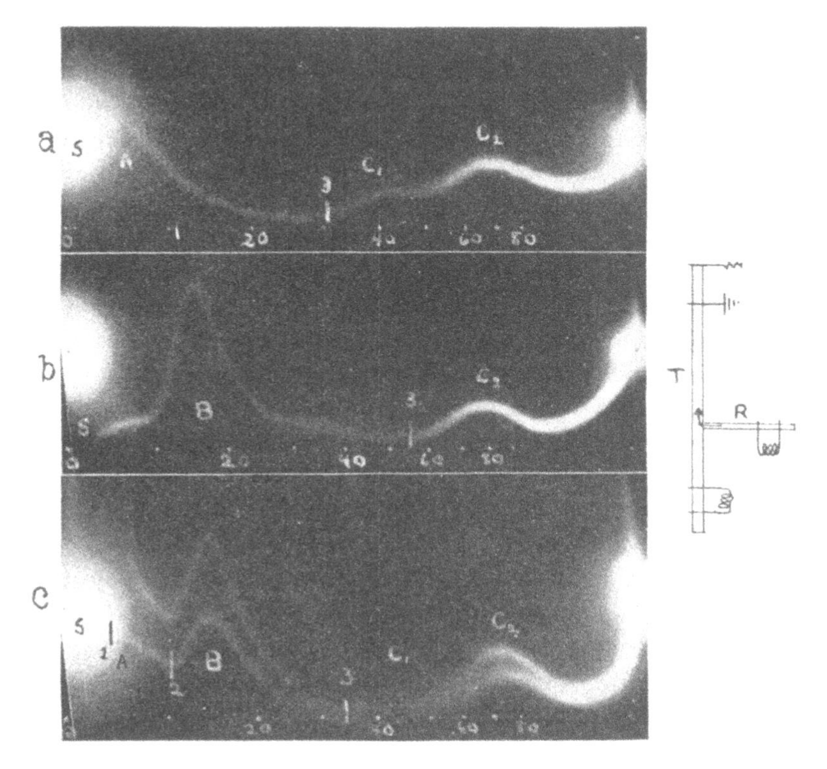

FIG. 31. The contribution of a gray ramus to the action potential of a sciatic trunk, bullfrog.

a, trunk stimulated; *b*, ramus stimulated; *c*, trunk and ramus stimulated simultaneously (two sweeps).

rangements are made to stimulate with induction shocks either the central end of the trunk or the free end of the ramus. For record *a* the central end of the trunk, and therefore all of the fibers supplied directly by the two roots, were stimulated. The record shows an A elevation and a double C elevation, C_1 traveling at the rate of 0.7 m.p.s., and C_2, 0.44 m.p.s. For record *b* the ramus was stimulated. A shock escape, *S*, can be seen at the start of the record, but no A elevation; and for the first time in this series of experiments we find a B elevation; it is traveling at the rate of about 3 m.p.s., a rate that is comparatively slow, but nevertheless within the range of B in the frog's sciatic nerve. There is also a C_2 elevation. Then, laying the free end of the gray ramus on the electrodes touching the trunk, both trunk and ramus were stimulated simultaneously. All of the elements that enter into the make-up of mixed nerve, namely, motor root, sensory root, and gray ramus fibers, thus are brought into action. In the resulting action potential, record *c* (two sweeps), are to be seen all of the components of the action potential of mixed nerve, an A element, a B element, and a double C wave.

The histology of the gray rami in relation to the problem of the sources of their fibers is an extremely obscure subject, and this is particularly true of the frog where the white and and the gray rami often are not separable structures. Our own histological controls have not been complete enough to justify any definite conclusions. Since, however, conduction rate in our experience seems to vary as the square of the diameter, we venture a prediction, based on the B conduction rate, relative to the size of the fibers in the gray ramus which produce the B elevation. When the A conduction rate has been 42 m.p.s., we have observed a B conduction rate in the sciatic of about 4.2 m.p.s. On this basis the effective B fibers of largest size should have a diameter, roughly, of 5.8 *microns*.

Experiments similar to those just described have been carried out in the cat and dog, also. The gray rami of these animals are not so readily prepared as are those of the bullfrog, and perhaps it is for this reason that the results of the experiments have not been consistent. All of the preparations successfully made have given good C elevations, such as is seen in record *a* of Fig. 32 (conduction rate of about 1 m.p.s.). Relatively few, but all of them with leads on the femoral nerve, have exhibited, in addition, an elevation with its threshold and rate of conduction definitely within the B range. Records from one of the latter are shown as *b* and *c* of Fig. 32. The sweep speeds for these were adjusted so as to emphasize B in record *b*, and C in record *c*. The conduction rate of the B elevation was about 10.5 m.p.s., and of C, 1.6 m.p.s. The velocity ratio of B to C of 6.5 is about the same as that, 6.2, obtaining in exemplars from the frog. Four of the preparations, all with leads from the sciatic plexus, were without B waves. The explanation of such inconstant results in mammals must await an opportunity to control the electrical analysis by histological analysis.

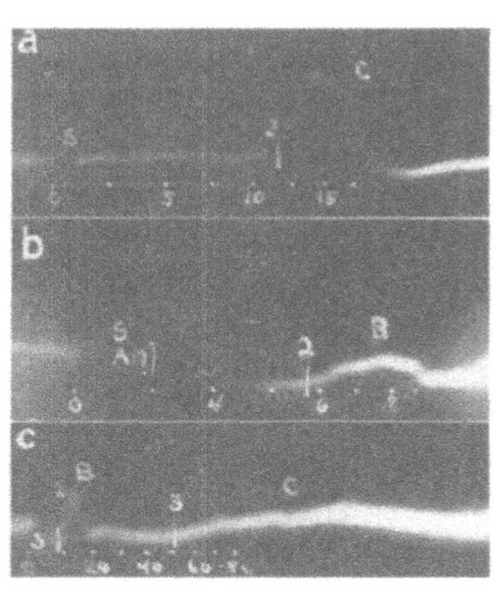

FIG. 32. Gray ramus action potentials, mammals.

a, from an excised gray ramus; the only elevation is C (1 m.p.s.).

b and *c*, gray ramus stimulated, lead from femoral; there is a B (10 m.p.s.) and a C (1.6 m.p.s.) elevation.

At no time during the course of this consideration of the sources of fibers have we found it necessary to regard as characteristically different the fibers supplied via the several paths, with the possible exception of those producing the C elevation of motor roots. There are on record, however, some observations which are regarded as incompatible with the concept of a continuous nerve fiber spectrum. To refer to just one, the statement is made (O'Leary, Heinbecker, and Bishop, 1934) that in the depressor nerve of the rabbit "autonomic myelinated fibers and the smaller afferent depressor (cerebrospinal) fibers are indistinguishable anatomically by present methods, but differ in their physiological characteristics"; that though the sizes of the largest fibers of both sources are "almost alike," the depressor ("5 *micron*") fibers conduct "about twice" as fast as the sympathetic ("4 *micron*") fibers. Just how the rate for 5 *micron* fibers was acquired is not made clear, since 5.5 *microns* is the specified size of the largest "depressor" fibers. If, under the circumstances, we arbitrarily take 5.5 and 4 *microns* as the respective sizes then, on the basis of a linear relationship between conduction rate and fiber diameter, the derived conduction rate ratio is 1 : 1.4 rather than the 1 : 2 relationship that actually obtains. This difference would be sufficiently significant to justify the point made by the above-mentioned authors. If, however, conduction rate varies as the square of the fiber's diameter, and we have supplied evidence indicating that it does, the conduction rate ratio becomes not 1 over 1.4, but 1 over 1.89. Determined by the latter calculation the conduction rate of depressor fibers does become "about twice" that of the autonomic fibers.

Relation of Fiber Grouping to Function

We come now to a consideration of the significance of the peculiar conglomerations of the constituent fibers of a nerve which give to conducted action potentials their characteristic configurations. It is so striking a phenomenon that it very naturally excited our curiosity immediately upon its disclosure by the electron oscillograph. From the very first it was surmised that the features are the expression of a segregation of fibers into different functional systems. Considerable plausibility was lent that guess by the fact that the sensory fibers, at least in the A range, form a series of elevations whereas the motor fibers form only one; for since there are several sense modalities and only one kind of motion, it was natural to suppose that on the sensory side each elevation is made by the fibers of a sense.

The first attempt to put this working hypothesis to the test seemed to confirm it (Erlanger, 1927). Records were obtained from a cutaneous nerve and from a nerve to muscle under exactly comparable conditions. The preparation employed consisted of the excised femoral nerve of the dog with two branches attached, one, the saphenous, a cutaneous, nerve, the other a nerve to one of the muscles of the thigh. The parent trunk was stimulated and the resulting action potentials were led from the branches separately. In Fig. 33 the records from the muscle nerve are on the right. Records *e* and *f* exhibit, at two amplifications, a simple A elevation traveling at the rate of 90 m.p.s. Record *g*, made with a slower sweep, brings out an insignificant, slower elevation, *B*. And a further slowing of the recording speed in *h* is needed to show a C elevation. The saphenous action potential, on the left, presents (in records *a* and *b*) an A elevation, traveling at the rate of 73 m.p.s., indistinctly composed of two waves, labeled β and γ, a very prominent B elevation (seen in records *b* and *c*) traveling 20.0 m.p.s., and also some C elevations (record *d*). As the labeling shows, and this was based on the relative conduction rates of the several elevations, the A elevation of the muscle nerve seemed to consist of *alpha*, and of the saphenous nerve of *beta* and *gamma*. Since *alpha*, *beta*, and *gamma* are features of the sensory roots and *alpha* a feature of the motor roots, the absence of *alpha* in the cutaneous nerve and its presence in the muscle nerve was regarded as evidence indicating that *alpha* of the sensory root action potential is made of a pile of proprioceptive fibers, and that fibers mediating cutaneous sensations do not contribute to *alpha;* that *beta* and *gamma* and *delta*, the last labeled *B* here, had to do with cutaneous sensations. Furthermore, since *delta* (or B) was made of the slowest fibers known at that time, it was assumed that nonmedullated fibers contributed their potentials to it. And, finally, since it had been concluded by Ranson and Billingsley (1916) that small medullated and nonmedullated fibers of the posterior roots mediate the sense of pain, the *delta* elevation was assumed to delimit the pain fibers.

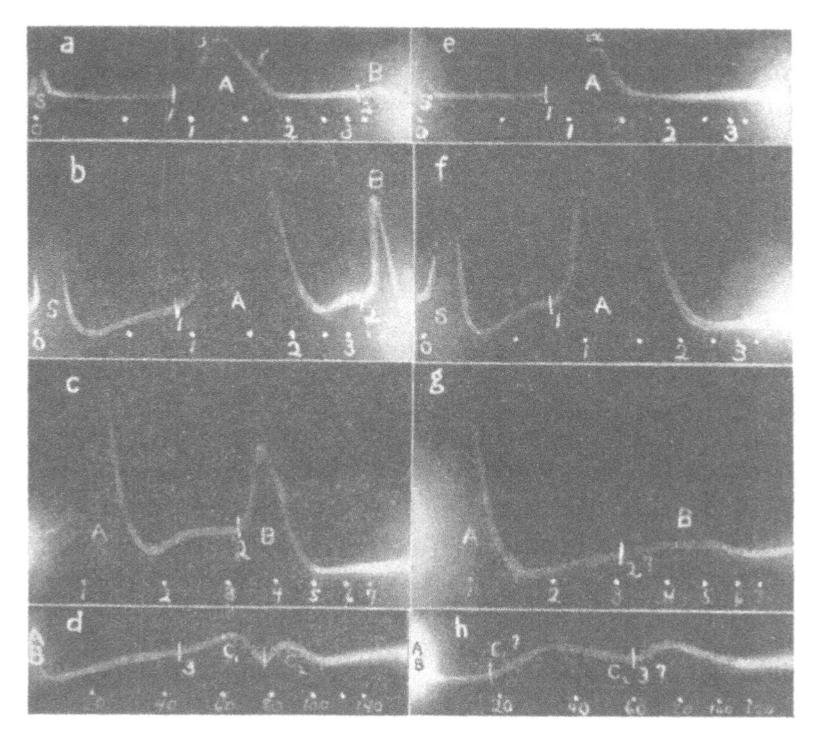

FIG. 33. The femoral nerve is stimulated. *a*, *b*, *c*, and *d*—lead from cutaneous (saphenous) branch; *e*, *f*, *g*, and *h*—lead from muscle branch; conduction distances 6.95 cm. *a* and *e*, low amplification; all others, high amplification. Time in msec. as indicated.

It becomes necessary to add a word of explanation here regarding the relation of *delta* to B. The designation *delta*, it will be recalled, was abandoned when high amplification failed to disclose a *delta elevation* in the sciatic action potential of the frog. Subsequently, the elevation added to the sciatic of the frog by the gray ramus was designated B. Although, as has been explained, we were unable to demonstrate to our entire satisfaction that B of the saphenous nerve of mammals is derived from the sympathetic system (Erlanger and Gasser, 1930), the fact that there is neither in the record from the sensory roots nor in that from the sciatic nerve of mammals (or frogs) an *elevation* sufficiently pronounced to be comparable with the beautiful B of the saphenous action potential, induced us to conclude that this saphenous B is sympathetic in origin. We expected to find in the sensory roots more than a "definite potential maximum" as the source of B. More recently Heinbecker, O'Leary, and Bishop (1933) have conclusively shown through degeneration experiments that the saphenous B elevation is in fact made by sensory fibers. Therefore, if alphabetical designations are to be continued, saphenous B should be renamed *delta*. The fact that the *delta* region of the sciatic and of sensory roots does not bulge, whereas in the saphenous it presents a fine elevation, presumably is to be attributed to the diversion into the saphenous nerve of large numbers of fibers of the *delta* sizes (of the *delta* sense, if you will).

Now, while it probably is true that agglomeration of fibers of a size has a functional basis, evidence has been accumulating which seems to indicate that the limits of any functional group are very much wider than are those of a fiber hillock. That, for example, is the conclusion to the story that is told by the exploration of skin nerves of the frog. The extent of the spread is shown in a particularly instructive manner in some experiments by Harris (1935). In the brain-pithed frog it is possible to elicit characteristic reflexes through tactile stimulation of selected skin areas. Thus a light touch applied to the skin over the plantar surface of the foot with a wet camel's hair brush, elicits a pure *extension* of the leg of the same side; but similar stimulation of the dorsum of the foot results in ipsilateral *flexion* of the leg instead of extension. The nerves going to these regions of the skin can be dissected free and can be stimulated electrically. Stimulation of the one likewise evokes extension, and of the other, flexion. Therefore, by stimulation of these nerves, one can determine the excitability ranges of the fibers, the more excitable of which certainly are touch fibers, eliciting in the one case an extensor reflex and in the other a flexor reflex. Then, to ascertain the position occupied by these fibers in the nerve spectrum, one can compare their excitabilities with that of the most irritable fiber in the sciatic nerve. One finds in this way that the extensor (afferent) fibers have at one extreme of their range an excitability, and therefore a conduction rate, that is quite as high as that of the fastest of the motor fibers. They contribute to *alpha*. This, it will be recalled, is not the case in the dog's saphenous nerve, whose fastest fibers have not the *alpha*, but rather the *beta* rate. At the other extreme of this extensor reflex effect are fibers with roughly one-half of the maximum excitability. On this basis one concludes that the fibers that produce reflex extension, presumably touch fibers, have conduction rates that range between say 40 and 14 m.p.s. Such fibers would contribute to three of the elevations, namely, *alpha*, *beta*, and *gamma*. But the fibers that elicit the flexion reflex, also touch fibers, at least in their upper ranges, collected, however, in another nerve, have a very different range. Their conduction rates, estimated by the same method, probably lie between 27 and 6 m.p.s., a range which would include the very slowest of the *alpha* fibers, the *beta* and the *gamma* fibers and on down to the fastest of the fibers contributing to B. One cannot be sure that the less irritable fibers concerned with these reflexes are tactile in function. There is evidence, however, that touch fibers can conduct as slowly as 6 m.p.s. For instance, there are to be found in the frog small skin nerves which contain no fibers conducting faster than that rate (Blair and Erlanger, 1933). Such nerves must mediate touch impulses.

Similarly, the fibers concerned with pain cover a size range which extends probably through the *delta* and the C ranges, and are both medullated and nonmedullated (Clark, Hughes, and Gasser, 1935).

It has been maintained (O'Leary, Heinbecker, and Bishop, 1935, p. 647) that "unit functional modalities are usually rather strictly confined to fiber-size groups that appear in the analysis of mixed nerves," and that "the usual range of such groups [is] 1 : 2 or 3." The latter statement signifies that if the largest fiber of a group measured 18 *microns*, the range of a "group" would include all elevations down either to *delta* or to B. But there are reasons for believing that the 1 : 3 range even cannot be wide enough to include the fibers of a function. That range, for example, is not wide enough to care for the range in fiber size that can result from the division of a fiber in a nerve trunk. For we have seen that the branches of a fiber may differ in size by a ratio that probably is as high as 1 : 4, a difference that could carry the function mediated by those fibers from the *alpha* quite into the B range.

The functions mediated by autonomic fibers will, of course, be confined to fibers conducting at rates that lie within autonomic limits which, in the frog, practically do not exceed 4.5 m.p.s. Personally we have had relatively little experience with this phase of nerve physiology, but it would appear that in this system, likewise, the fibers of a given function are not sharply confined to the limits of recognizable elevations in action potentials. To give just one example, inotropic action exerted on the auricle of the turtle's heart through the vagus nerve develops when the stimulus attains a strength that elicits "the first sign" of the elevation in the action potential that has been designated B_3 (Heinbecker, 1931); but the inotropic effect continues to increase as the C fibers come into action

(Heinbecker and Bishop, 1936). In this case, therefore, we are dealing with an autonomic function that is mediated by fibers contributing to two elevations, one of medullated, the other of nonmedullated origin, presumably.

In contemplating the material that has been brought together in this lecture the feature that most excites one's curiosity is the diversity of the sizes of nerve fibers and the even greater, but associated, diversity of conduction rates. One naturally wonders whether the variety that has been disclosed can have any significance relative to the functioning of the nervous system as a whole. In the case of voluntary motor fibers the teleological significance of the size differences is not immediately apparent. It is perhaps justifiable to infer, however, that the larger fibers innervate the quicker muscles and the larger motor units. Moreover, if the discharge from the central nervous system is synchronized (Gasser and Newcomer, 1921) the differences in conduction time to a muscle along different fibers might be a factor in the production of a steady tension when, during feeble contractions, the rate of stimulation of the units is below that necessary to produce complete tetanus. Thus, if the distance of conduction were, say, 0.5 meter, impulses starting simultaneously from the cord could, because of velocity differences, arrive in the muscle separated by an interval of over 0.3 sec. Therefore, the different units of a muscle even during the feeblest efforts could be in every stage of contraction.

On the afferent side one might suppose that the size of the fiber carrying an impulse into the central nervous system determines the magnitude of the effect delivered and also that a specific combination of a variety of intensities is responsible in part for the local sign of peripheral stimulation. The fact that the diameters of the peripheral and central projections of the fibers of the dorsal root ganglion cells differ does not preclude the possibility that fiber size is of some significance in this respect. It is necessary to realize in this connection, however, that each of the branches of the terminal arborization of a fiber in the central nervous system is going to exhibit properties characteristic of its own size, not those of the axon from which it springs. Therefore, it is impossible to tell from the size of a peripheral fiber what its central characteristics are going to be. All that we can assume is that large fibers will probably have extensive arborizations and produce correspondingly large or extensive central effects.

In addition, the variety of fiber sizes, through the associated range of impulse velocities, might have a great deal to do with the process of integration within the spinal cord. Gasser (1935*a*) has suggested in this connection that impulses carried into the central nervous system by the fastest fibers, which would include proprioceptive and tactile fibers, might "set the neural connections which determine the nature of the reflex which is to follow" when, some time later, the impulses arrive which are conveyed by the more slowly conducting fibers. To illustrate by an example he employs, if you should stub your toes impulses would arrive in the cord via the muscle-sense fibers, and possibly via tactile fibers also, in less than 0.02 sec. But impulses would continue to arrive via the smaller fibers for as long as 2 sec. after the impact, the last impulses to be delivered, probably being those eliciting pain. And all the while they would be falling into a nervous system conditioned by the impulses first to arrive there. These, however, are subjects that we know so little about that we must desist from carrying them beyond this very indefinite suggestion. If the variety of fiber size has a functional significance it remains for the future to determine just what it accomplishes.

We have through this lecture held to a particular alphabetical designation of the elevations exhibited by the conducted action potentials of nerves. Now, if fibers differ from one another only by virtue of size (this certainly is true of medullated cerebrospinal fibers, and probably true of all fibers) the original tentative system of alphabetical labeling obviously becomes meaningless. Indeed, it can be shown that such a classification is apt to give an impression of a definiteness which does not in fact obtain.

The designations do not delimit fiber types; neither do they necessarily designate fibers of the same size ranges when they are applied to apparently homologous elevations in different nerves; and since a considerable fraction of the fibers divide as they approach the periphery it is obvious that the constitution of a nerve relative to the sizes of its fibers must be constantly changing along its length. The letters, moreover, do not serve to delimit functional groups of fibers, since unit functions usually, perhaps invariably, are mediated by fibers contributing their potentials to more than one of the elevations, though functional grouping of fibers with respect to size probably is responsible for the action potential pictures. All of these assertions could be supported with experimental evidence were there time. There is just one point that might be expanded. The letter C designates the last elevation, or group of elevations, of the mixed nerve action potential; it is supposed, also, to stand for nonmedullated fibers (Bishop and Heinbecker, 1930), those, for instance, of the sensory roots. Now there are some animals whose sensory roots do not contain nonmedullated fibers, for the reason, apparently, that all of the fibers become large enough to acquire medullary sheaths. This is true of *Bos*, for instance (Duncan, l.c.). Electrograms of bovine dorsal roots still have to be made. One can predict, however, that the last elevation will be found to correspond in its relative position, with C in other animals. If so, this would constitute a case in which an elevation is made by fibers which do not belong in the C category according to one definition, but which mediate the functions cared for by so-called C fibers of other species.

It is time, then, that a logical system of designating fibers be inaugurated. Such a system would define fibers either by their conduction rates, or by their sizes, or by both conduction rate and fiber size, should that prove to be necessary. If, at the same time, the temperature were supplied at which observations are made, it would be possible to include in one system all of the fibers of all of the vertebrates we know about, and so to reduce all fibers to common terms.

In order to indicate how such a system would work let us get a specific case clearly in mind. Let us apply the scheme to the case of the fiber supply of a mixed nerve. A rate of 100 will be assigned the fastest fiber. For many reasons this is a convenient value to take, not the least important of which is that 100 m.p.s. very commonly approximates the conduction rate of the fastest fiber in warm-blooded animals, probably including man, at body temperature. In addition, it permits us to express conduction rate on a percentile basis. The maximum fiber size for the scheme might be put at 18 *microns*. This, roughly, is a median value for the largest fiber in vertebrates. The diagram of Fig. 34 shows to the left one pair of spinal roots, the white ramus and the gray ramus, and to the right the fully constituted mixed nerve. Included in the diagram is a table showing in *round numbers* the conduction rate ranges of the fibers of the principal elevations; this will serve to indicate where they fall in this system. The continuous range of rates in the sensory root is 100 to 0.5, in the voluntary motor outflow 100 to 2, in the sympathetic outflow 2.0 to 0.5, and in the inflow from the sympathetic, 10 to 0.5. The start of the elevations in the action potentials of these structures would be described by the following numerals: *alpha*, 100; *beta*, 60; *gamma*, 40; *delta* (if there were one in the root) 25; B, 10; and C, 2. Where the diameter of the largest fiber of a species happens to be significantly smaller than 18 *microns*, in the monkey, for example, (Bishop, Heinbecker, and O'Leary, 1932) a reference value to replace the 100 in the present system could be derived by a calculation based on the proportionality of conduction rate to the square of the diameter. And, finally, to care for temperature differences, as in the case of cold-blooded animals, cross reference would involve, in addition to any necessary adjustment needed to care for a difference in the size of the largest fiber, a correction for any temperature divergence there happened to be from 37.5°. The latter

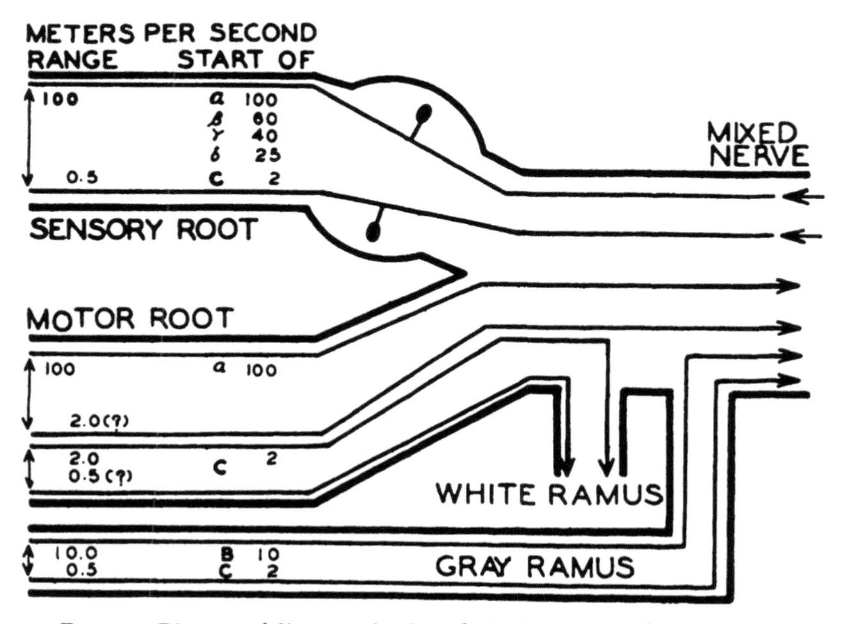

FIG. 34. Diagram of fiber constitution of mixed nerves on the basis of a conduction rate of 100 in the largest fiber.

adjustment could readily be made, since the temperature coefficient of conduction rate within usual temperature ranges remains relatively constant at 1.7 (Gasser, 1931). Features on the action potential of any nerve are fully described, therefore, when their conduction velocities are given along with the size of the largest fiber of the species and the temperature at which the observation is made, and this is the method of designation we are advocating.

III

SOME REACTIONS OF NERVE FIBERS TO ELECTRICAL STIMULATION

THROUGH a period of about ten years following the introduction of the electron oscillograph into physiology, or up to the year 1932, our observations were confined to responses amplified about 100,000 times (effective amplification about 200 microvolts = 7.5 mm.). Under these circumstances the successive records obtained from a preparation with conditions constant are one just like the other even when the stimulus is threshold and produces a just visible response, despite the fact that, under such circumstances, a just visible deflection is made up of the fused responses of a number of axons. When higher effective amplification (about 10 microvolts = 7.5 mm.) became available we were at first rather disconcerted to find that the successive threshold-conducted responses of large nerves, instead of being superimposable, as they always had been, were extremely variable in appearance; the variation in fact was kaleidoscopic in character, as may be seen in Fig. 35. Records 2 to 8 were made in succession under conditions which seemed to be perfectly constant, yet they have every conceivable shape. Of this variability, only a small part is attributable to artifact, as may be ascertained by comparison with record 9, which shows the noise level of the apparatus; it is a sweep like the others in this figure, but made without stimulating the nerve. Neither can the variability be attributed to irregularities in the stimulating circuit. Record 1 demonstrates that. It is a half maximum *alpha* spike from the same preparation, on a slightly slower time line, and amplified considerably less, so that the discrete effects of single axons are below the limit of visibility. Record 1, moreover, is made up of 40 superimposed successive sweeps. Had the successive shocks varied by as much as 0.25 per cent such superimposability as is seen here would have been impossible.

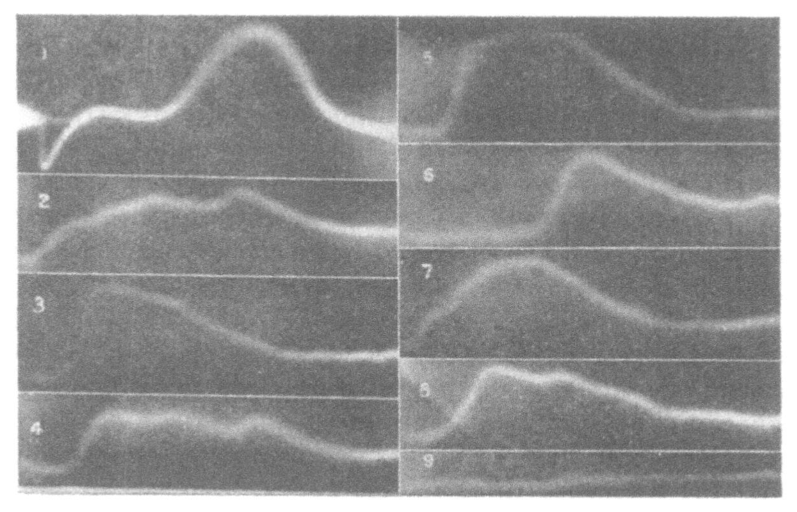

FIG. 35. Variability of multifiber threshold responses.
1. Half maximum *alpha*, moderate amplification; 40 superimposed sweeps.
2–8. Successive single sweeps, higher amplification; threshold stimulation.
9. Blank sweep, to show noise level in *2–8*.

Shock-Response Delay and its Variations

As a matter of fact, it can readily be demonstrated that the variability of these pictures is due to a spontaneous and independent variation of the excitability of the individual responding fibers. That such is the case became apparent when we (Blair and Erlanger, 1933) began to make records of the action potentials of individual axons from a slender preparation with the higher amplification. The preparation employed was the phalangeal nerve of the frog, to which we have referred so often in the previous lectures. Occasionally, in such a preparation, the two most excitable axons happen to have about the same excitabilities and the same conduction rates, but axon spikes with distinctive markings of one kind or another, one, for instance, that is more diphasic than the other. Under such circumstances it can be seen that successive responses of the two axons to threshold shocks come and go and shift to and fro in a wholly unpredictable fashion. Both fibers may respond to one shock, one only to the next, the other only to a third, or neither axon may respond, and in so doing the responses are constantly shifting their relative positions. Such behavior cannot be ascribed to variation in the stimulus, for then the two fibers would vary alike as the stimulus changed; only an independent variability fits the premises.

The width of the fluctuation in excitability can be measured by the change in the ratio of the number of responses to the number of stimulations as the strength of the shock is increased in steps. It is convenient in making these measurements to stimulate regularly, at the rate of, say, 30 shocks per minute. As the shock strength is increased from a subthreshold level a strength will be reached at which the fiber responds only rarely, say about once in 20 stimulations. To increase this low probability of response to a high probability, where, for example, the fiber *fails* to respond to a shock in the same ratio, i.e., 1 in 20, it usually is necessary to increase the strength of the shock roughly about 10 per cent.

How, in the premises, is the threshold stimulus of a fiber to be defined? Should we take as the measure of threshold the strength of current that induces one response, or that fails to stimulate once, in the course of an arbitrary number of successive stimulations? Obviously there would be practical difficulties in the way of applying either of these criteria; the determination of the threshold in either case would become an exceedingly tedious process. So instead we have defined the threshold of a fiber as the strength of current which stimulates in approximately half of the applications. This is what we have meant in previous lectures by the term "threshold" when it was used in connection with the stimulation of single nerve fibers. The criterion still is somewhat arbitrary, and still is tedious in its application; for, to be certain that the stimulation strength is such that the fiber responds in about 50 per cent of the trials obviously requires that a large number of trials be made.

The *shift* in the position of the successive responses on the screen, it can be shown, is due to fluctuation of the interval elapsing between the brief stimulating shock and the initiation of the fiber's characteristic response; there is, in other words, a fluctuating shock-response delay. That it is this that is determining the fluctuation and not variation in the rate of propagation of the impulse can be demonstrated by recording the responses where they originate at the cathode of the stimulating shock (Blair and Erlanger, 1936*a*). It is not, however, a simple matter to obtain legible records under these circumstances since one of the terminals of the stimulating current connects directly with the amplifier and, at the high amplification needed to record the responses of single fibers, the shock escape badly deforms the record. Line tracings of records made in this way, and demonstrating a shift, are seen in Fig. 36. Here three sweeps are superimposed, two (*A* and *B*) showing spikes developing in different positions on the shock artifact and the third (*C*) the shock artifact alone as recorded when, through a spontaneous fluctuation in the excitability of the fiber, the shock failed to elicit a spike. This response shift, amounting to 0.17 msec., was not the widest given by this axon. The longest shock-response intervals and also

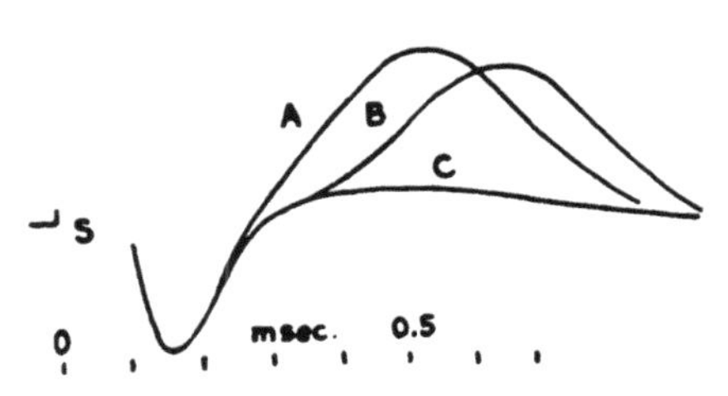

FIG. 36. Spontaneous fluctuation in shock-response time in an axon, under threshold stimulation. *S–C*, shock artifact; *A* and *B*, responses.

the widest shifts are seen when the shocks are just strong enough to elicit an occasional response; the shock-response intervals then may range between 0.2 and 0.4 msec. The intervals, it is interesting to note, may be as wide as the time to maximum of a spike. If the shock strength is increased the range becomes narrower and the shock-response interval shorter, until, with shocks about 10 per cent above threshold, the shift becomes vanishingly small, and the shock-response interval then may be something less than 0.1 msec. It may be added that this, namely 0.06 to 0.1 msec., is about the interval that obtains regularly when the unconducted response of a large nerve to a shock is recorded with moderate amplification (Erlanger and Gasser, 1924; Bishop, 1927). The longer intervals, displayed when dealing with single fibers, are not seen under the latter circumstances because the shock is well over the threshold of the most irritable fibers when the deflection becomes appreciable.

Students of physiology are familiar with the term 'latent period' as applied to the response of tissues to electrical stimulation. As usually employed the term refers to any interval there may be between the termination of a brief applied stimulus and the characteristic response of the tissue. Now we have just seen that axons, when stimulated by an induction shock which lasts, say, 0.04 msec., can respond after an interval as long as 0.4 msec., and the question arises, is this shock-response time a latency properly so called? There are reasons for believing that it is not. If we regard the firing-off process as a trigger mechanism, latency might be defined as the period during which the hammer is falling. Such time as is put into pulling the trigger to the point where the hammer is released would then not be a part of latency. Now if we stimulate a fiber with a threshold shock and then, after this shock has run its course, but before the fiber has fired off, pass another shock through the same electrodes in the opposite direction, we find that it is possible to prevent the first shock from taking effect (see also Dittler, 1925). The hammer, in other words, has not yet begun to fall. Whether this can be accomplished through the entire shock-response period is not quite certain. Possibly there is preceding the response an interval of less than 0.1 msec. during which the counter shock is without avail; this may respresent the time required for the hammer to fall. If there is a latent period as we have defined it, we can say then, that it does not exceed 0.1 msec.

It will be explained more in detail below that conceivably these brief shocks, when threshold, stimulate after their termination because of the electrical characteristics of nerve, which possibly are referable in part to the membranes interposed between the potential source and the irritable axon. It will be suggested that due to the brevity, and consequently the high intensity of the potential necessary to stimulate, the membranes are immediately charged to a level exceeding that needed to set off the irritable mechanism, and that then the charge held up by the membranes is transferred to the irritable mechanism during and after the period of the shock to bring the axon to the firing-off point after the applied voltage has fallen to zero. To put the matter in another way, the applied potential seems to be distorted by the physical properties of the nerve in such a way that the effective potential is attained only after a delay. The local exciting process initiated by the stimulating shock and continuing after it apparently is not, like the characteristic response of the axon, a reaction, which, once initiated, develops independently of the imposed conditions.

Though the excitability of fibers is subject to fluctuations, conduction rate is not. This is proved by the fact that the conducted axon spikes initiated by a sufficiently strong shock, a shock strong enough to eliminate fluctuation of the response time when the record is from the stimulated locus, all arrive at the lead after exactly the same interval. It might, nevertheless, be more discrete to say that, if there are spontaneous fluctuations in conduction rate, they are local, and that there are so many loci conducting faster than the average rate as there are loci conducting more slowly than the average. However this may be, the fact that the conduction rate is constant makes it unnecessary in routine determinations of shock-response variations to obtain the record directly from the stimulated locus; such fluctuations as occur in the time of arrival of the spike at the lead are assignable to fluctuations in shock-response time at the site of stimulation. Shock-response *time*, however, can be determined only by leading directly from the stimulated locus.

Constant Current, Response Time Relationship

It has been seen that a threshold *shock* induces responses irregularly and after a fluctuating shock-response interval. Similar, but very much wider spontaneous fluctuations in

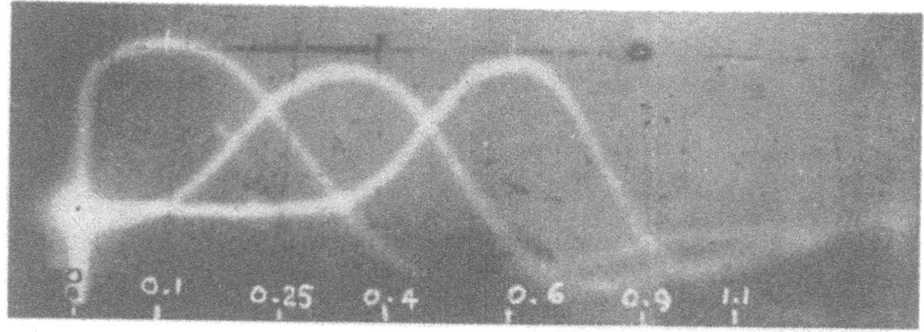

FIG. 37. Spontaneous temporal shift of the responses of an axon to three successive applications of a threshold constant current.

response time are observed when for stimulation a rectangular *constant current* is employed in place of a shock (Blair and Erlanger, 1933 and 1936*b*). The shifts then, at room temperature, may be as wide as 1.5 msec. They are, moreover, much more easily demonstrable because the threshold voltages of the constant currents are very much lower than those of shocks and consequently distort the base line less. This may be seen in Fig. 37 which shows the spontaneous shift in the response time of an axon in three successive applications of the same "threshold" constant current. Here the shift in crest time is of the order of 0.5 msec. The time intervening between the make of the

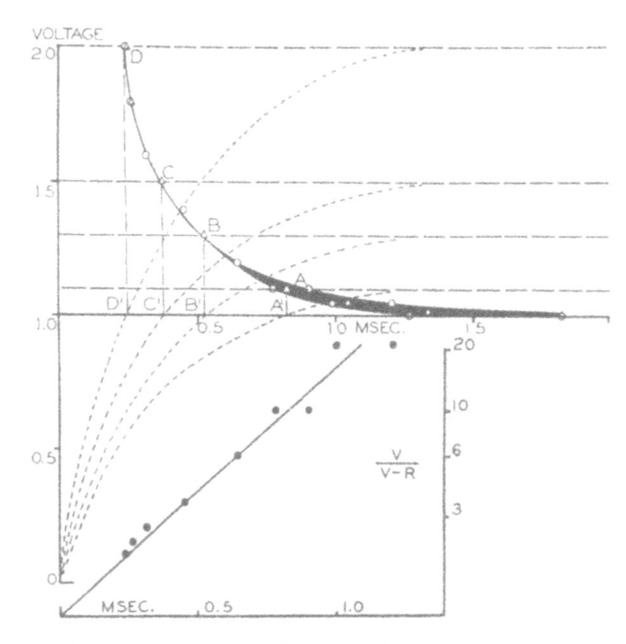

FIG. 38. Upper graph (circles): Strength of constant current (V), 1.0 being rheobase (R), plotted against the response time of an axon.

Lower graph: Ordinates are $\frac{V}{V-R}$, abscissae time; all values derived from upper graph.

threshold constant current and the response of the axon does not show in this picture. A usual range with rheobasic currents is between 1 and 1.5 msec. but it may be very much wider. To be specific, the range at threshold is indicated in Fig. 38 by the circles on ordinate 1.0; they show that the position of the spike shifts between 1.26 and

1.83 msec., zero time being the start of the rectangular constant current. Increasing the strength of the constant current causes both the make-response time and the response time range to diminish. The next two circles above and paralleling those representing the shift at threshold show the make-response time range associated with the indicated higher level of the stimulating constant current. By following upwards the course of this curve as marked by the circles it can be seen, to repeat, that as the current strength increases, the spontaneous range of the response time diminishes; it diminishes at such a rate that when the current strength attains about 1.3 times the threshold, the shift has become insignificant. In the meanwhile the make-response time also has been diminishing. The solid curve, therefore, expresses the relation which obtains between the make of a constant current of any strength and the time the fiber responds; it is a make-response time curve.

The interpretation of this curve and the discussion of its significance can be simplified by making certain assumptions regarding the manner in which a nerve is stimulated by an electrical current. Let us suppose, as we have done before, that a fiber fires off when the voltage acting on the responding mechanism attains a critical value, and that the structures surrounding the mechanism, including the membranes of the fiber itself, act like a condenser that is charged by the applied current through a resistance. Then, without going into any of the details, which are too complicated for consideration here, the strength-response time curve should fit the formula which underlies the usual mathematical treatment of strength-duration relations:

$$V = \frac{R}{1 - e^{-\frac{t}{k}}} \text{ where}$$

V is the applied voltage,
R the rheobase and
t the time.

Now mathematicians tell us that a curve defined by this formula should plot a straight line when the logarithm of $\frac{V}{V - R}$ is plotted against the time. Therefore, the values represented by the circles in the upper curve of Fig. 38 have been plotted in this manner in the lower graph, $\frac{V}{V - R}$ against the time. The result is indeed a straight line—a straight line, moreover, that passes through zero. One can conclude with some assurance, therefore, that nerve responds to an applied current as does a system such as is shown in Fig. 39, which explains itself. We wish to emphasize, however, that we do not regard this as a replica of the irritable mechanism of nerve, but only as a schematic representation of the changes in the excited state that occur in nerve during the initial stages of the action on it of an applied potential.

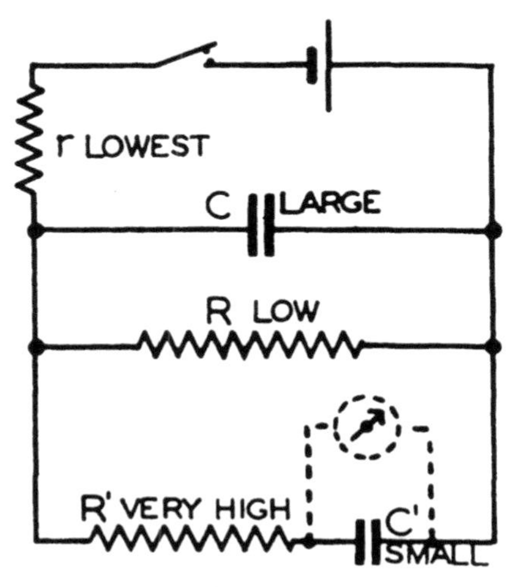

FIG. 39. A circuit in which the potential would rise on the make as it does in nerve.

Now how are we to account for the response time ranges which are plotted in the curve of Fig. 38? In attempting to answer this question we call attention first to the fact that the excitability of a fiber is constantly varying, and through a range, you will recall, of about 10 per cent. In the diagram of Fig. 40 this spontaneous variation in threshold is represented by the band labeled "Spontaneous threshold variation." Now if the excitation determined by a rectangular current of threshold strength rose along a curve resembling the lowest one in this diagram, it is obvious that when the momentary threshold of the fiber is at its higher limit the excitation process would have to rise to the limit in order to stimulate and then the fiber would respond late, at *1*; whereas when the threshold of the moment happened to be at its lower limit the fiber would respond at the earlier point, *2*. The response with succes-

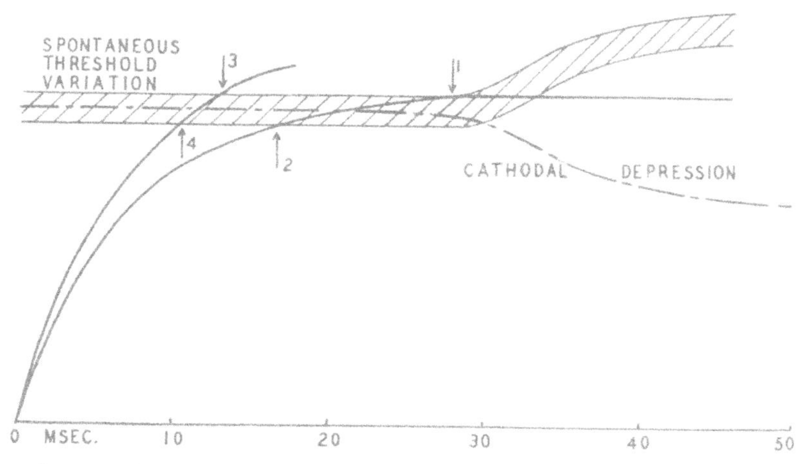

FIG. 40. Diagram illustrating the relation between the width of the spontaneous threshold variation (the band) in a fiber and the response time shift when the stimulus is the make of a constant current. The shift when the current is rheobasic is between *1* and *2*; under a stronger current, between *3* and *4*.

sive stimulations would then range between *2* and *1*. Increasing the strength of the constant current so that excitation rose along the upper of the two curves obviously would cause the responses to range between *3* and *4*, and so on. The response time shifts are readily accounted for in this manner.

Excitation and Depression by Constant Currents

As a measure of the excitability of a point on a nerve one commonly employs the intensity of current needed to stimulate. That testing current is said to have threshold strength which is just strong enough to produce the least perceptible response. As the testing current any form can be used, shocks, condenser charges, constant currents, etc. When dealing with relatively large nerves, such as the sciatic or the saphenous, and moderate amplification, the spike of a single axon is below the limits of visibility; threshold responses under such circumstances are made up of the summed spikes of a considerable number of the most excitable axons. The nerve, for this reason, appears to be a stable structure, always responding in the same manner, apparently, to the same stimulus. Moreover, when dealing with multifiber responses one can get a measure of the variations in excitability much more conveniently and quickly merely by determining the *variations* in the height of responses to a submaximal testing shock of constant strength, or by determining the strength of shock needed to keep the response at a selected submaximal level. For the moment, therefore, let us depart from single axon responses and describe, as they occur in large nerves, the more reproducible changes in excitability developing at the cathode in association with the closure of a constant current through the nerve. The graphs of Fig. 41 (Erlanger and Blair, 1931*b*) all show that following the make of the constant current, excitability at first rises rapidly but eventually reaches a plateau which effectively is level. After something like 1.5 to 2 msec., at 12° C., in a very much shorter time at higher temperatures, the plateau is ended abruptly by a curve which declines most rapidly where it starts. The graphs give the impression that we are dealing with two processes starting in succession each at maximum speed. The first process must be the excitation occurring at the cathode of the constant current and the second cathodal depression or what has been termed "accommodation."

The first determinations of excitability curves of this character demonstrating a momentary rise above the steady level finally attained (Bishop, 1928),[1] yielded a shape that seemed to match that formed by the counter-electromotive forces of two condenser-systems in series; the curves exhibited no such discontinuity as is described above. The decline from the maximum of excitability was not regarded

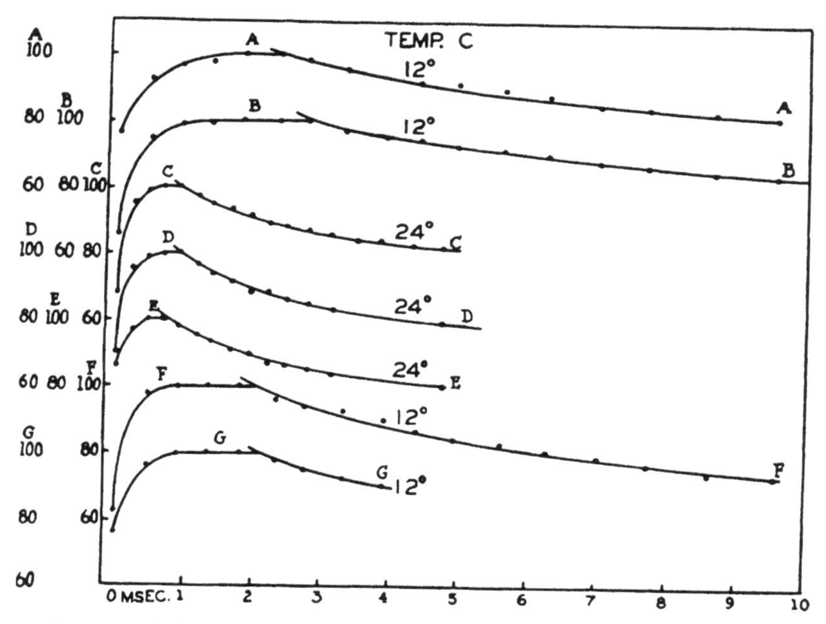

FIG. 41. Excitability variation with the time following the make of a constant current, 90 per cent of rheobase. Multifiber responses. Constant height method. Ordinates, percent of maximum excitability. Abscissae, time in msec. Showing the effect of temperature.

as a manifestation of cathodal depression, for it was stated that cathodal depression appears late and only with strong currents. It can be shown, however, (Blair and Erlanger, 1936*b*) that the decline from maximum excitability, when measured, say, 20 msec. after the end of the plateau, proceeds progressively at all strengths as the polarizing

[1] Monnier (1934, p. 9) attributes the original finding to Tigerstedt (1882). Tigerstedt, however, finds the decline insignificant or absent, and fails to make anything of it.

current is increased. The decline gives no evidence of the initiation of a new process when the polarizing current exceeds a certain strength.

The excitability curve, we now know, has such a shape as to signify that excitation and depression are two unrelated processes—that one process succeeds the other; either this, or if they start simultaneously, that the curve of depression is sigmoid in shape, beginning very slowly with the make of the current at zero time, and acquiring maximum velocity very much later, as shown in Fig. 40 by the broken curve. A process which might be expected to proceed in such a fashion, is one that is determined by the diffusion of cations through the nerve membrane under the influence of the polarizing current. The delay in the appearance of cathodal depression could then be attributed to the time required for the ions to traverse the membrane, the rapid increase of depression which then begins, to the appearance of the ions on the outside where they exert their effects, and the slow attainment to the maximum of depression, to a correspondingly slow attainment to a new equilibrium state under the influence of the polarization.

But now we must confess that even these more recent curves seem to be distorted pictures of the rise of excitation and of the onset of depression. That this may be the case became obvious when we began to deal with single axons. In such experiments it can be seen that there is a distortion which is due to the fact that with the change in excitability determined by the constant current there occur wide variations in the shock-response intervals and in the response play. Fig. 42 illustrates the variations; it pictures the result obtained when, by the shock-test method, the threshold of a *single fiber* is determined at the cathode of a rectangular constant current that is somewhat weaker than the rheobase. The curve indicated by the dots expresses the variation in threshold with the time following the make of the constant current. It is the same curve as the series of excitability curves shown in Fig. 41 excepting that it is plotted in terms of threshold, not in terms of excitability, so that the plateau now is down, not up. The feature of present interest is the play of the shock-induced spikes. The response-time range under the action of the constant current when just threshold is shown by the separation

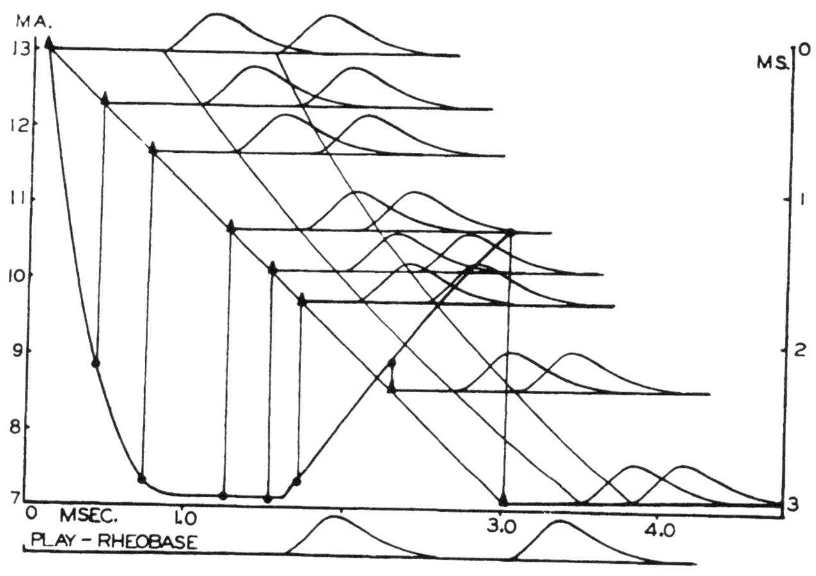

Fig. 42. A graph (dots) of the change in threshold of an axon produced by a constant current 90 per cent of rheobase, as tested by induction shocks. The conventionalized spikes labeled "play-rheobase" indicate the rheobasic response time and its play, the other pairs of spikes the shock-response time and its play at indicated points (the dots) on the curve.

of the conventionalized spikes on the line labeled "Play-Rheobase." The other pairs of conventionalized spikes show the response-time play when threshold shocks are delivered at the cathode of the subthreshold constant current at the intervals after the make of the constant current which are indicated by the vertical lines connecting with the curve the lines on which the spikes are drawn. It is perfectly obvious that the play changes as the polarization proceeds.

It is necessary to bear in mind in this connection that the position of the end of the plateau is not altered materially by changing the strength of the constant current. This fact is illustrated in Fig. 43, in which are reproduced excitability curves derived under several polarization strengths ranging from 50 percent of the rheobase in *A*, to

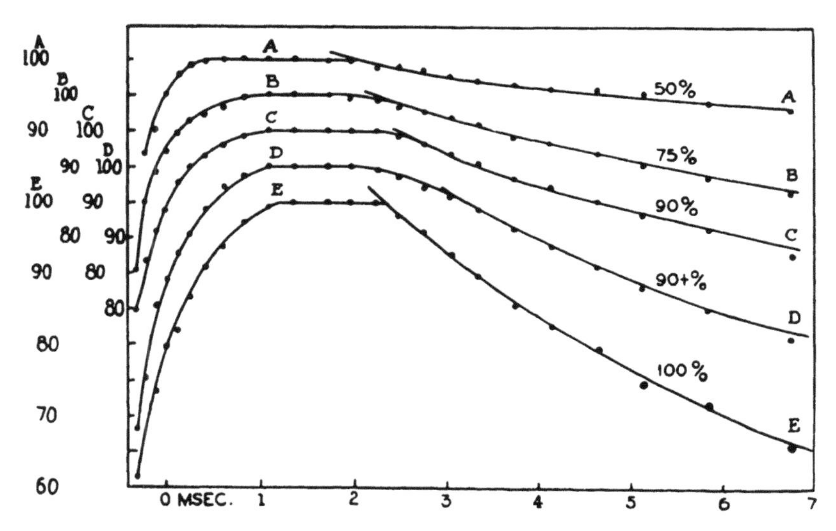

FIG. 43. Excitability variation with the time following the makes of constant currents of the strengths (in terms of rheobase) indicated. Multifiber responses, constant height method. Ordinates, per cent of maximum excitability.

100 per cent in *E*. The plateau ends at the same time in all. Now it will be noted in Fig. 42 that the spike play under rheobasic stimulation is all later than the angle marking the end of the plateau: the responses appear to be developing while threshold is rising, when cathodal depression apparently has raised materially the fiber's threshold. Since this seems incredible, we are forced by the picture to infer (*1st*) that the curve of thresholds given by the shock-test method is distorted, and presumably by the

fact that the shock effect sums with the constant current effect in such a way as to cause the responses to materialize long after the delivery of the shock and (*2nd*) that cathodal depression probably becomes significant after an interval following the make of the constant current that reaches out to, or almost to, the limits of the play of the responses elicited by rheobasic stimulation, in this case after 3 msec.

The picture of the changes in excitability occurring during a rectangular current as derived through the shock test method, is inaccurate, due, we believe, to the fact that we are measuring the excitability produced by a current of one configuration (a rectangular constant current) in terms of a current of another configuration, namely, an induction shock. A more adequate method of gauging the time of onset of depression, it would seem, would be to determine the behavior of the responses when a slightly subrheobasic constant current is brought to threshold by the addition to it (after different intervals) of another *constant current*. Then the excitation curves of both the conditioning and the testing currents would be the same.[2] An experiment employing such a step-like stimulus is illustrated by Fig. 44. The temperature in this case, it should be noted, was unusually low (10° C.) and the time values are correspondingly long. The play of the responses under rheobasic stimulation is again shown on the line labeled "Rheobasic Play." It is between 3.5 and 7 msec. The increase in

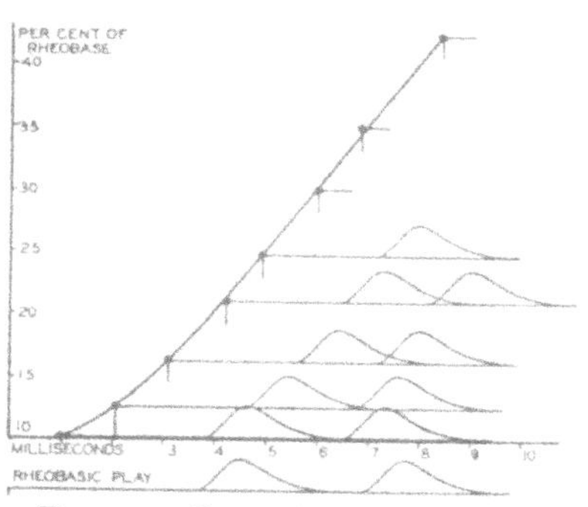

Fig. 44. Curve (dots) of the decline in effectiveness of a 90 per cent rheobasic constant current, starting at 0 time, in terms of a second constant current applied *en échelon*. The conventionalized spikes show the several response plays; uppermost pair not complete.

[2] Hill (1936), however, believes this procedure introduces other complications.

the strength of the testing steps needed to make them threshold as the makes of the steps are moved later in the conditioning subrheobasic current is given by the curve passing through the dots. That curve does not interest us particularly now, but the play of the responses at the several levels does. This play again is indicated by the pairs of conventionalized spikes. Of the several significant happenings which they picture, it is necessary for present purposes to note but one, namely, that the spikes in their play do not begin to fall later than the play at rheobase (7 msec.) until a considerable interval separates the make and testing steps. Such behavior is taken to signify that it is in the neighborhood of 7 msec. that depression intervenes and for a time prevents responses from developing later as the exciting steps move later. There must, of course, be a limit to this effect, but the limit is not shown in Fig. 44. These interpretations, it may be reiterated, are based on the view that the range of play of the spikes is determined by the slope of the part of the excitability curve upon which operates the spontaneous 10 per cent vacillation of the fiber's excitability.

We now are prepared to compare the strength-response time curve, which we have derived through our own experiments, with the curve which long has been employed by physiologists as the graphic expression of the law of stimulation of nerve by constant currents, the so-called strength-duration curve. The latter is determined by plotting the strength of a constant current against its threshold duration. At every strength of constant current the duration of the current is shortened until it becomes just threshold. In Fig. 45 such a curve is plotted (dots) from data derived from an axon. In the same figure has been plotted a strength-response time curve (circles) derived from the same axon under exactly comparable conditions. The curves obviously are different. At only one point do they coincide, and that is where the current is rheobasic. In other words, when the current is cut off at what is termed the principal utilization time the response follows after an inappreciable interval. At all higher voltages the strength-duration curve falls inside of the strength-response time curve—the response follows the constant current break after an interval. What happens at the high end of the strength-response time curve we have not been able to determine, since it has not been possible to carry the observations on an axon through the last 15 per cent of the time.

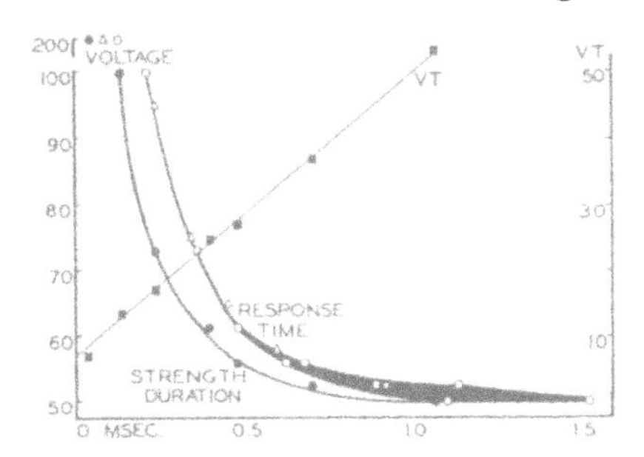

Fig. 45. Curves showing the relation (*a*) between current strength and duration (dots), and (*b*) between current strength and response time (circles). The triangles show the calculated response time values. The *Vt* curve (squares) indicates that the strength-duration curve has the usual hyperbolic configuration.

Since there is a break-response time it is obvious that applied rectangular currents can continue to act after their termination; therefore the time of effective current action, if measured to the break, is too short. It is this fact (possibly there are others) which causes the classical curve to depart from the simple condenser curve. The deviation has not, however, been ascribed to that circumstance, but rather to the onset of accommodation. Our experiments, however, indicate, as has been seen, that accommodation does not become appreciable until there has elapsed a period which is about as long as the utilization period and that it, therefore, does not materially modify the configuration of the first phase of strength-duration curves. Finally, the fact that make-response time curves, through the range in which they can be determined (about 85 per cent of the utilization time), fit a simple law, in turn confirms our belief that the onset of cathodal depression either is immediate, but very gradual at first, or is delayed and sudden.

The Rise and Fall of Excitation by Brief Currents

Frequent reference has been made to the fact that the spontaneous variation in the excitability of a fiber at the cathode of a stimulating current is equivalent to about 10 per cent of the voltage necessary to excite, and we have just seen how that fact can be employed in accounting for the shift in make-response time when a constant current is used to stimulate an axon. Now, the same phenomenon provides a means of gaining information relative to the rise and fall of excitation that is determined by brief currents such, for example, as an induction shock. It tells us that the excitation of the fiber, starting from the normal or zero level at the time a threshold shock is delivered (Fig. 46, lower diagram, dotted curve), rises to reach a crest long after the shock (S_1) is over; that, as a matter of fact, the crest is attained somewhere between 0.2 and 0.4 msec. later; that at 0.2 msec. (at *a*) the rise crosses the ordinate lying 10 per cent below the crest; and that the fall crosses that ordinate at 0.4 msec. (at *b*). We might, on this basis, expect the excitation curve initiated by the shock to have a temporal configuration resembling that of the dotted curve, *1* + *2*.

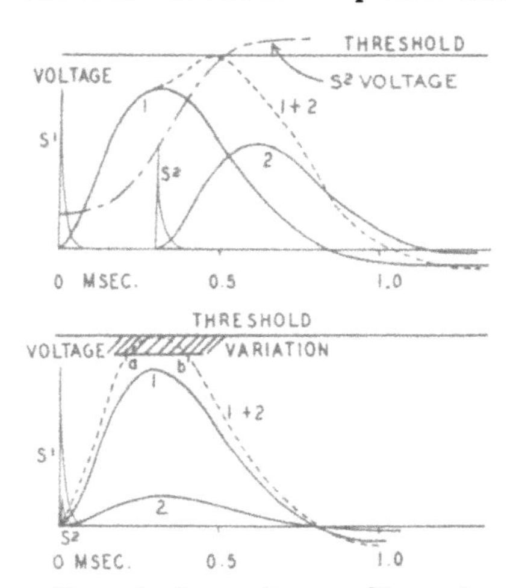

FIG. 46. Lower diagram: Illustrating the range (*a* to *b*) of the shock-response time play determined by the interaction of excitation (dotted curve) and spontaneous threshold variations.

Upper diagram: Illustrating the method of plotting the change in threshold determined by a subthreshold shock (S').

It is possible, however, to obtain experimental data which tell us something more about this process. The method of procedure has been referred to casually in connection with our determinations of the summation intervals of different fibers (Lecture II). The nerve is conditioned by a relatively strong shock (S_1, upper diagram), one having, say, 90 per cent of the threshold voltage. This conditioning shock is followed by a testing shock (S_2) and in each temporal position the latter is given the voltage necessary to bring it to threshold strength. One then finds that when the shocks are superimposed, as in the lower diagram, the testing shock stimulates when it has a voltage equal to 10 per cent of threshold. But in order to keep the testing shock (S_2) at threshold value as it is moved away from the conditioning shock (S_1) its voltage must be increased, and as indicated by the curve labeled "S_2 voltage" in the upper diagram. To keep the testing shock threshold in the usual case, it must be given a strength of 100 per cent of threshold, that is to say, the excitability of the fiber becomes normal, when the interval between shocks attains a value lying somewhere between 0.2 and 0.6 msec. (in the case of this diagram, at 0.5 msec. where the S_2 curve crosses threshold). This is the summation interval. But the end is not yet. The excitability of the fiber as affected by the conditioning shock continues to fall so that the testing shock as it moves still later must be given a strength that *exceeds* the normal threshold of the nerve. This depression of excitability reaches its maximum usually in about 1 msec. and then the excitability returns to normal during the course of about 3 to 4 msec. Some curves illustrating these changes in excitability of nerve following a submaximal shock are shown in Fig. 47. The technique employed here differs, however, from that just described in that the gauge is not the shock strength but, in effect, the height of the multifiber response resulting from a shock of constant strength. The figure shows (Erlanger and Blair, 1931*a*) that the excitability at first falls rapidly as the shocks separate, to reach the normal level, the 100 per cent level, in from 0.2 to 0.6 msec.; that then the fall continues on down to a subnormal level before excitability begins to return to normal.

This *depression interval* sometimes fails to materialize, and much remains to be done in the direction of ascertaining the conditions which determine its variability. A beginning, though, has been made. Graham (1935*a*), for example, has found that one of the conditions it is necessary to take into consideration is the composition of the medium bathing the nerve. With the particular Ringer's solution she regards as normal, the maximum depression attained rarely exceeds 1 to 2 per cent. An excess of potassium increases both the depth of the depression and its duration, whereas increased amounts of calcium, if considerable, tend to decrease, or may actually eliminate, the depression.

Since the amount of depression developing depends upon conditions which are independent of the applied voltage, it follows that when depression is in evidence the duration of the summation interval is of questionable value as an index to the reactivity of tissues. Therefore there is no reason for being concerned about the fact, referred to in the second lecture, that the relation between the summation intervals of fibers and their velocities is not a simple one.

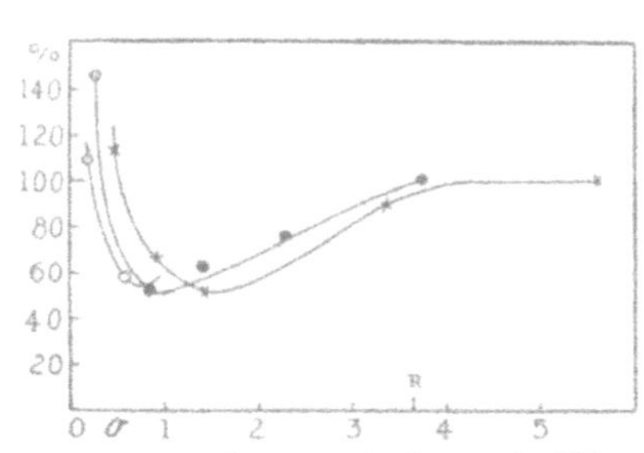

FIG. 47. Curves of the excitability changes induced by an induction shock; the normal level is 100 per cent.

The depression interval, however, is a phenomenon not entirely devoid of physiological significance. The uninfluenced depression interval, as may be seen in Fig. 47, has about the same duration as the relatively refractory period (indicated by *R*), the period required for the nerve's excitability to return to the normal level after it has actually fired off. In this particular case excitability was restored to normal in 3.7 msec. after both a subthreshold shock and a complete nerve discharge. This relation is seen more in detail in Fig. 48 in which are plotted both types of recovery as they occur under comparable conditions in the same preparation—the recovery from refractoriness (squares) according to the vertical scale as shown, and the recovery from the effect of a subthreshold sensitizing shock, on a very much magnified scale, not shown. All of the curves approach the normal level of excitability in approximately the same time, in this case about 3 msec. The correspondence suggests that the relatively refractory period may be related in some manner to the after-effects of the electrical phenomenon of nerve action.

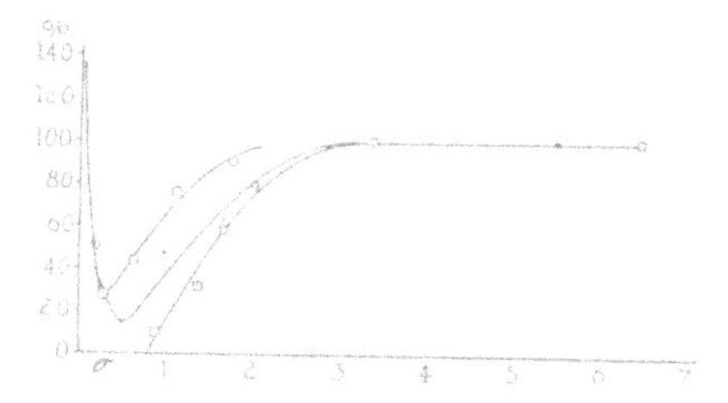

Fig. 48. Dots—graph of excitability changes induced by a brief shock just below threshold in strength; circles—by a much weaker shock. Squares—curve of relative refractoriness. The normal level is 100 per cent.

Another correlation of the depression interval that has physiological interest is with the so-called Wedensky phenomenon. Working with nerve-muscle preparations Wedensky found that a stimulus strong enough to elicit a contraction may fail to stimulate when it is repeated at certain relatively rapid rates. The many explanations of this phenomenon, which since have been proposed, have failed to carry conviction until Kato and collaborators (1929) ascribed it to an "after-effect," a brief period of lowered excitability left either by an ineffective nerve impulse or by an artificial stimulus of less than threshold value. If, as our observations seem to indicate, the relatively refractory period and the depression interval have essentially the same basis, the maintenance of the Wedensky block by a succession of nerve impulses seems simple enough to understand. It is quite obvious that the depression interval following a subthreshold shock can be maintained by repeating the shocks each one after an interval that puts it in the depression interval of a preceding one; that under such circumstances the second shock (and a third and so on) can be given a strength which would stimulate normally excitable nerve but fails to stimulate the depressed nerve, and instead maintains the depression. In the same sense, an action potential stimulating across a synapse or possibly also through an injured stretch of nerve would be followed by a relatively refractory period there; but should a second impulse, arriving at that time, fail to reach the threshold because of the relative refractoriness, excitability would rise and fall without reaching the threshold, and so with a third impulse, a fourth, and so on. Thus stimulation would fail to occur as long as the succession of impulses continued at an adequate rate. This state of affairs could not develop, it might be added, where the impulses are running in intact nerve, one in the relatively refractory phase of the other. It could not because under such circumstances the second impulse runs at a slower rate than the first, with the result that the former gradually lags behind the latter. As it does so, however, it gets into less refractory nerve and its rate of propagation picks up until, were the course sufficiently long, the second impulse eventually would become normal not only in its speed of propagation but in every other respect also (Gasser and Erlanger, 1925).

Repetition during Constant Current Flow

We come now to a subject, namely, axon repetition, which at first may seem to be wholly unrelated to what has gone before. We, however, are going to discover many points of contact, and particularly with the subject of excitability changes induced in nerve by constant currents. In the normal discharge of its function the tangible messages transmitted by a peripheral nerve fiber practically invariably consist of a longer or shorter series of rapidly repeated action potentials; yet there is evidence indicating that this intermittent activity of the fiber can be determined by a continuous liberation of energy in the sending station.[3] The clearest illustrative case, perhaps, is that of light reception: the retina, when steadily illuminated, develops in effect a continuous potential (Hartline, 1928), whereas the impulses transmitted to the brain by the optic fibers are intermittent (Hartline and Graham, 1932). To all appearances the sending stations initiate repetitive discharges along the fibers by causing a current to flow through them continuously. If such is the case, it should be possible, through the use of applied continuous potentials, to duplicate in nerve fibers the repetition pictures that are determined by end organs and by the central nervous system. Now, as a matter of fact, a constant current when closed through an excised nerve usually causes the fibers to respond (at the cathode) but once. If the current is made strong enough there may be another response, again usually a single one, immediately following its break, this response starting, however, from the anode. Sometimes under conditions which never have been clearly defined, closure of the current may induce a repetitive response, the so-called closing, or Pflüger's, tetanus; and even the breaking of the continuous current also may be followed by a repetitive discharge, the opening, or Ritter's, tetanus. These opening and closing tetani heretofore have been studied only in preparations giving multifiber responses. Under such circumstances, however, it is scarcely possible to ascertain what is going on since to a degree each fiber behaves individually and the total response, therefore, is complicated beyond comprehension.

During the course of our observations of the last two years on the behavior of individual axons we occasionally have encountered preparations which were suited to the study of the closing tetanus (Erlanger and Blair, 1936);

[3] The subject has been reviewed recently by Bronk (1935).

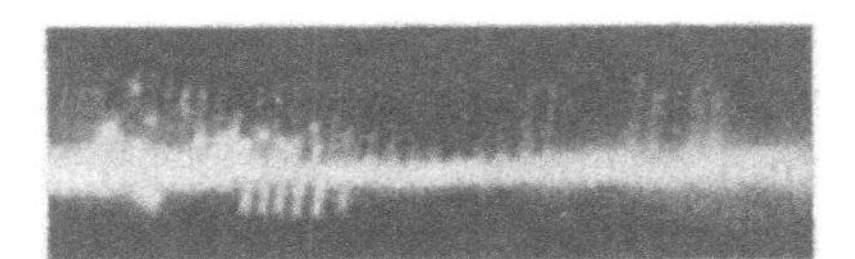

FIG. 49. Repetition induced by a rectangular constant current. Several axons are yielding monophasic, others diphasic, spikes.

repetition would be a better designation to use since it describes what happens in nerve as well as in muscle. We have employed for these observations the minute phalangeal preparation. It is, we repeat, the sciatic nerve of the frog extending down to include a branch supplying one of the terminal phalanges. The lead into the electron oscillograph is taken from this minute branch.

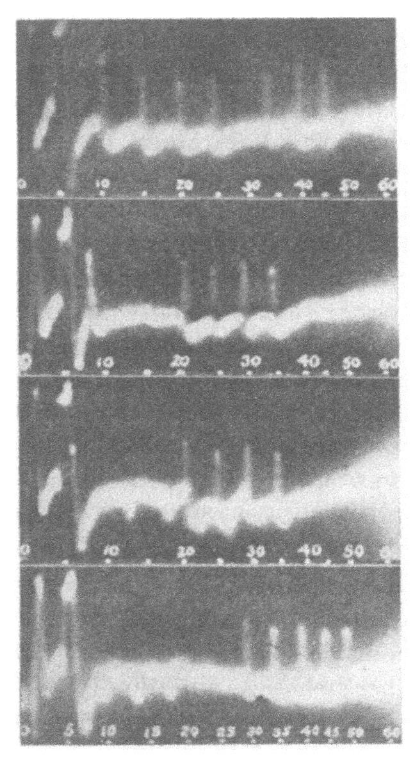

FIG. 50. Repetition induced by a rectangular current, conditions constant. A single axon is repeating in the two upper records, and two in the lower records. The deflection at zero time is the make artifact. Time in msec.

Relatively few of the preparations contain fibers which respond repetitively to the rectangular constant current. Moreover, one never can predict when repetition is going to occur. And even when repetition does develop, the different fibers of a preparation may in this respect behave in an entirely individual manner: stimulation at a given point on the nerve may cause some fibers to repeat, while others, perhaps equally or even more excitable by ordinary standards, cannot be made to repeat with any strength of stimulation. A rather characteristic picture from a preparation in which the rectangular current causes several fibers to repeat is seen in Fig. 49.

In general the repetition threshold is lower the larger the fiber, but small fibers may repeat when large ones do not. And those fibers of a nerve that do repeat may yield repetition patterns that differ widely one from the other. As may be seen in Figs. 49, 50, and 51, some of the fibers may start to repeat at once, or almost at once, when the circuit is closed, others only after a considerable delay. Delays of 20 to 80 msec. are perhaps commonest, but they have been longer even than 0.2 sec. The delays may vary rather widely in the same fiber from current application to current application, even when there has been no obvious change in conditions. On rare occasions the vacillation of the delay has ranged from 0 up to 20 to 30 msec. or more from trial to trial; an instance of vacillation is seen in Fig. 50. Immediate and delayed repetition apparently are not unrelated phenomena; one can change into the other, again without any obvious change in conditions.

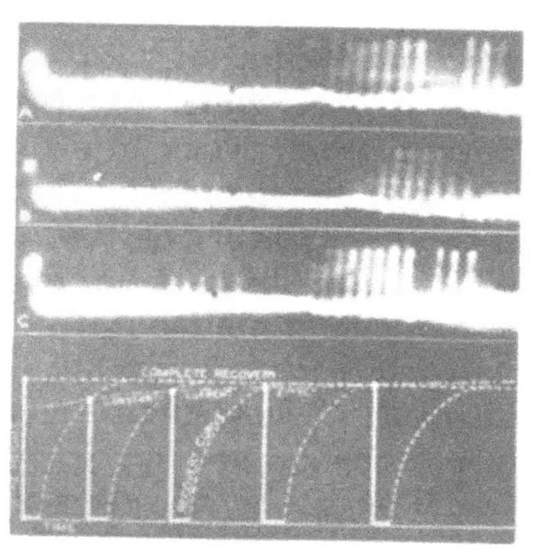

FIG. 51. Repetition induced by a constant current closed with the start of the sweep. The time is linear and is indicated by the following description of record *A*: Make to first spike, in msec., 92.5, and from spike to spike in succession, 4.26, 3.96, 4.26, 4.57, 5.03, 4.42, 15.4 (15.4 ÷ 3 = 5.1), 4.57, 5.50.

The diagram is an attempt to illustrate the basis of a repetitive burst.

When the rectangular constant current is applied at different points along the nerve it occasionally is found that one and the same fiber repeats wherever the stimulus is applied and with like patterns. More commonly, however, a fiber repeats at but one or two of the points tried and nowhere else. Again the different points collectively may present the entire gamut of responses.

The repetitive response of a fiber to a rectangular current often is made up of several successive bursts of discharges. The uppermost records of Figs. 50 and 51 show two bursts each. In such cases each burst has the general characteristics of a response that consists of but a single group. These characteristics, in the case of a short burst, are such as to give the impression that the response starts at a time when excitation is high but declines until the burst terminates. The diagram in Fig. 51 expresses the idea. It portrays in semiquantitative fashion an actual burst consisting of five responses like the burst of seven responses in record *A*. The spikes are represented by the heavy vertical lines; in height and spacing they show the amplitudes and time intervals of the spikes of the series. The first spike is the highest, the second the lowest, and the spikes then increase to almost attain the initial height when the series ends. At the same time the intervals between spikes gradually increase. In the diagram the recovery from each response, the relatively refractory period, is represented by a dotted curve. Now if the constant current produces an effect such as is indicated in the figure by the broken curve, the axon, as it recovered from each of the responses, would be stimulated at the points indicated by the dots, and the fiber would fail to respond when the constant current effect rose above the level of complete recovery.

It has been stated that a repetitive response of a fiber may consist of a succession of bursts. Now when the bursts are spaced by short gaps, such as are seen in the uppermost record of Fig. 50 and in records *A* and *C* of Fig. 51, they obviously consist of the absence of one or more of the spikes from a series which otherwise would have been perfect. In other words, the gaps have an aliquot relation to the spike intervals. One naturally would be inclined to attribute such gaps to the interposition of a block somewhere on the conducting pathway. But all of the evidence is opposed to block as the cause of the gaps. Thus there is little or no difference, in respect to the way they end, between a repetitive discharge consisting of but one burst, and the parts of a repetitive discharge consisting of several bursts. It is hard to believe that every repetitive discharge ends because of a growing block; therefore, it is unlikely that the intermissions are the result of block. This conclusion, moreover, is supported by experimental evidence. For instance, a spike initiated by a shock applied in a gap at any time outside of the absolutely refractory period of the last response invariably is conducted the length of the fiber. Since that propagated disturbance is not blocked, there is no reason for supposing that the constant current is initiating impulses which are being blocked. Again, when one and the same point on the nerve is both the lead and the cathode of the rectangular current that initiates the repetition, there is, during a gap, not the slightest sign of the initiation of spikes which fail of conduction. The potential of a spike initiated under these circumstances, even if the spike were unconducted, certainly would spread perceptibly to the lead. We are compelled, therefore, in accounting for the aliquot relation of the gap intervals to the spike intervals, to assume that the recovery period of a fiber from a response is the first of a continuing series of periodic oscillations in excitability with a period about equal to the recovery period of the fiber, and to consider the gaps as true intermissions of activity.

Some experimental evidence has been acquired indicating that there actually is such an oscillation in excitability. A description of one of the relevant experiments follows. A fiber that requires a superrheobasic current to yield immediate repetition is stimulated with a rectangular current that is above the rheobase and close to, but below, the repetition threshold, and the resulting excitability changes are mapped by the shock-test method during and following the recovery from the refractoriness following the make response. A typical result is plotted in Fig. 52. Following the recovery from absolute refractoriness the fiber's threshold falls to a steady state through three decrementing oscillations around a steady level, the low threshold points falling, it will be noted, at 5, 10, and 15 msec. The first low threshold point must be the equivalent of, if it is not actually, the supernormal phase. Be this

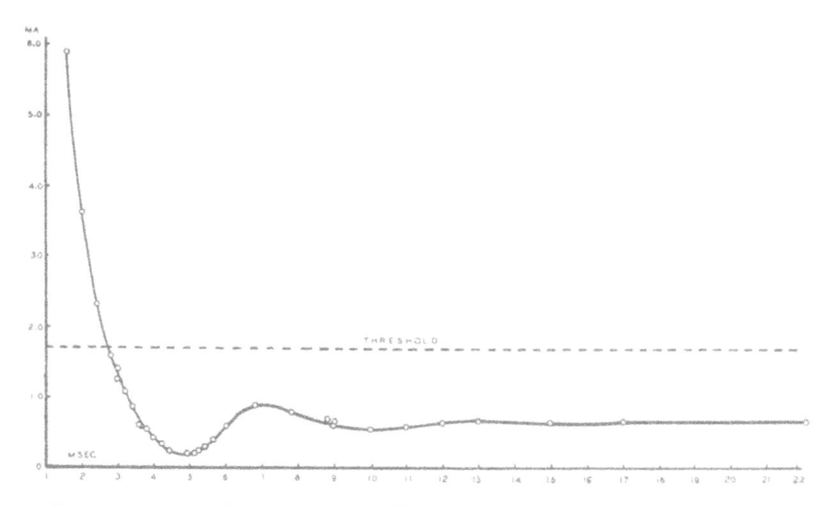

Fig. 52. Oscillation in threshold following the recovery of a fiber from the make response of a constant current which, at higher voltages, causes the fiber to repeat at a rate comparable to the oscillation rate seen here. The dotted line is the unpolarized *threshold*.

as it may, it must be more than a coincidence that the period of these decrementing oscillations, namely, about 5 msec., was of the same order of magnitude as the period of the repetitive discharges of this fiber. Therefore, even though the amplitude of the final excitability oscillation in experiments of this kind always has been low (possibly because it has been necessary to employ for activation a current strength lower than that which caused the repetition), we, nevertheless, are inclined to regard the oscillations as a manifestation of the mechanism that tends to continue the repetition period when, in a gap, the recovery incentive to repetition is lacking.

That a continuing oscillation in excitability is the basis of the aliquot relation between gap and repetition intervals is suggested also by the fact that one spike of a continuous series of repetition can be eliminated by means of a properly timed shock with anode at the cathode of the rectangular current that is inducing the repetition. That is to say, one of the responses of a series can be knocked out by this procedure without otherwise disturbing materially the sequence of the series.

I have said there is no telling how a given preparation is going to behave relative to repetition. Well, we had been endeavoring for some time to find a way of controlling the responses when eventually we happened upon some observations which pointed the way to the goal. Usually the general pattern of the repetition induced in a fiber by the application of the rectangular current at a given point remains true to type even if and as repetition gradually fails. On one occasion, however, the pattern changed suddenly and radically. The changes were of a kind that could have been produced only through action on the nerve of a change in a demarcation current. That demarcation currents are a factor in the production of repetition was suggested also by the comparative excitabilities of a fiber at the two loci in contact with the stimulating electrodes when from one, acting as cathode, the rectangular current elicits only delayed repetition and from the other, now acting as the cathode, either an initial response only or immediate repetition. This has by no means been a rare combination. In the majority of such instances the one of the two loci that yielded the delayed repetition had the lower excitability; occasionally the difference was as great as 3 is to 1. Such a difference, it was realized, could result from anodal polarization of the nerve in the region of low excitability. In the light of these indications that demarcation currents affect the repetition picture and that the repeating point is anodally polarized it was natural to suppose that the conditioning factor for repetition is *anodal* polarization.

But before describing how this inference was followed up I want to consider for a moment still another clue which was happened upon while investigating the excitability changes produced by the passage of a rectangular current. Earlier in this lecture a description was given the threshold pattern determined in a fiber by a rectangular current at its cathode. You will recall that the threshold, as in curve *A* of Fig. 53, falls at first to a plateau, H, and then, after about 1 to 2 msec., rises again. A fiber that repeats *after a delay* yields, we found, a very different threshold curve. We have seen curves of the types that are illustrated by both *B* and *C* and various combinations of these. There may or may not be an initial plateau; in either case a low threshold point is reached late, usually in about 30 msec. Such curves, let me repeat, are obtained from fibers yielding delayed repetition. From a *fiber yielding immediate repetition* the curve of thresholds seen in Fig. 52 was obtained. This fiber repeated under a current of three rheobases, but the determinations upon which this curve is based were made when the rectangular current was reduced to the point where it just failed to produce repetition but still had greater than rheobasic strength. It conse-

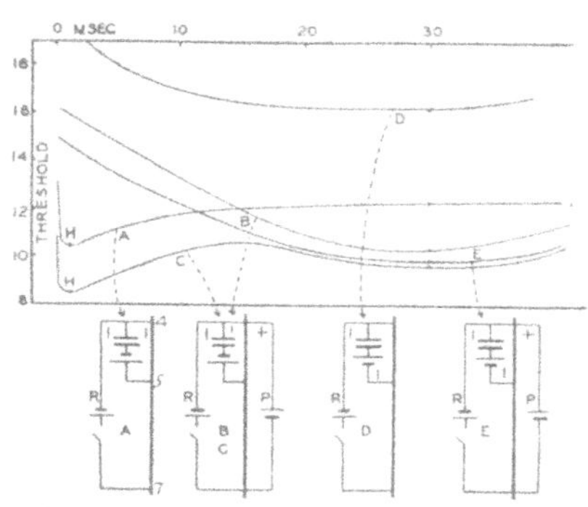

FIG. 53. Type curves (semidiagrammatic) of threshold changes in axons at the cathode of a rectangular current, with diagrams indicating the conditions under which the several types develop. *H* is the plateau region.

quently elicited a make response, and the curve, you will recall, shows the recovery from the refractoriness that follows the make response and in addition some subsequent oscillations to which reference has already been made. The significant behavior from the present standpoint is not these features, but rather the failure of cathodal depression to develop. The usual eventual rise in threshold is absent here. I would like also to call attention at this time, for use later on, to the fact that this curve, and all similar curves obtained under the same conditions, show a transient phase of supernormality at the peak of recovery.

We find, then, that a fiber which responds to a constant current with an action potential at closure has a threshold curve that falls immediately to a plateau and then rises; that a fiber which responds with immediate repetition yields a threshold curve which gives no evidence of cathodal depression—the threshold falls quickly to a low level and remains there; and finally that when the repetition develops late the curve of threshold falls slowly to a late minimum and then rises slowly.

Now it was not surprising to discover that the threshold curve of a fiber accords with the type of response the fiber yields. The observation of real and present interest is one that came of an experiment that was indicated by the observations suggesting that a repeating fiber is anodally polarized. It was found that it is possible to convert a threshold curve of the configuration of *A* of Fig. 53 into one shaped like *B* or *C* by imposing the cathode of the stimulating rectangular current on a locus that is being continuously polarized anodally; and by the same means to convert immediate repetition induced by a rectangular current into delayed repetition and to increase the duration of the repetitive response. The effect is shown in Fig. 54. The rectangular constant current starts in each record at *M*. Record *A* is the response to the rectangular current alone; the repetition is immediate, or practically so. The other records were made by applying the rectangular current while the nerve was being continusously polarized anodally with strengths increasing from *B* to *E*. The increasing delay in the start of the repetition is obvious.

Now, since an applied continuous anodal polarization conditions the nerve for a repetitive response at the cathode of a rectangular current, it seems justifiable to conclude that repetition, or at least delayed repetition, induced by a

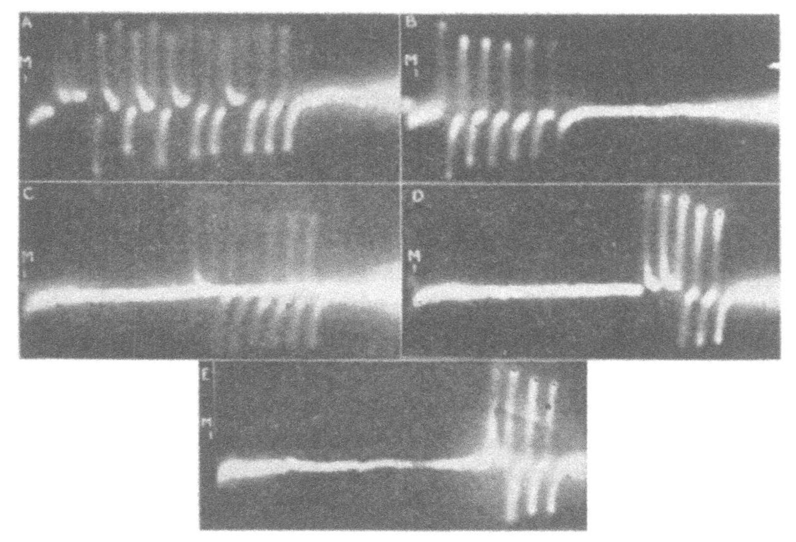

FIG. 54. Repetition in an axon induced at the cathode by a rectangular current, illustrating the effect of continuous anodal polarization. The latter increases in the successive records from *B* to *E*, inclusive. The time is linear; record *E* subtends 72 msec. from *M* to the end.

rectangular constant current in nerve not polarized anodally by artificial means, develops when the nerve is anodally polarized by its own demarcation potential. Intermittence, it may be added, is most commonly observed in delayed repetition, and since delayed repetition is par excellence the result of anodal polarization, it would seem that intermittence likewise is associated in some manner with anodal polarization.

The rate of the repetition we have been describing is fast; it could scarcely be otherwise, if restimulation is the result of recovery from refractoriness. As a matter of fact, however, even the shortest repetition interval of a burst always is longer than the relatively refractory period as directly determined. Typical values from a preparation are 5.8 msec. for the first (and the shortest) repeater interval, and 4.0 msec. for the relatively refractory period. This difference is too wide to be assignable to error. The restimulation, therefore, is not due to simple recovery from refractoriness. There is, however, another correlation which probably is significant. We have seen, in the case of a fiber that repeats immediately, that the repetition interval and the time to supernormality following the recovery from a response are of the same order of magnitude (see Fig. 52). Now it is obvious that recovery via supernormality would be much more apt to elicit a response than return to normal without supernormality. And it is of interest to add that in the course of some experiments, still unpublished, Blair has found that anodal polarization converts recovery without supernormality into recovery with supernormality. Undoubtedly, the tendency to induce supernormal recovery is another of the effects attributable to anodal polarization that make for repetition.

Thus far we have been dealing, as we have just said, with fast repetition; the rates usually have been in the vicinity of 200 per sec. But physiological repetition is rarely, if ever, as fast as this, and may be as slow even as 10 per sec. Such slow repetition we have encountered only twice in our long series of observations on excised nerves. Typical pictures obtained from one of the preparations are seen in Fig. 55. The time in these records falls off logarithmically. The rectangular current begins in each record at *M*; it is relatively weak in record *A* and increases in the successive records. Two, perhaps three, fibers are repeating. One, producing the series labeled with numerals, is a large fiber;

it repeats slowly, at threshold, beginning with the make of the current. Another, a small fiber producing the series labeled with letters, repeats rapidly at its threshold, reached in record *F*. It is not necessary to give additional details since there has been relatively little opportunity to study repetition of the slow type. The only observations of any

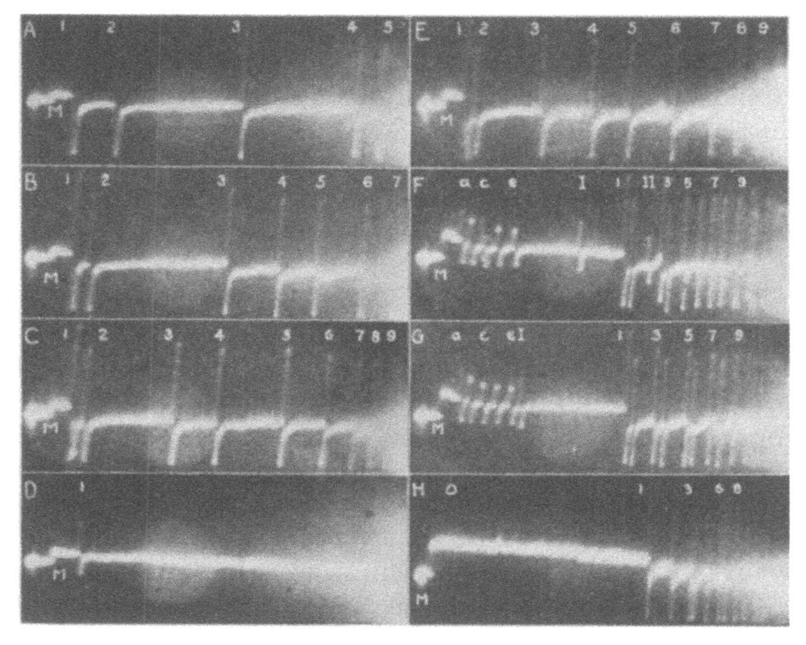

FIG. 55. Slow (and fast) repetition in axons induced at the cathode of a rectangular current successively increased in strength from *A* to *H*. The time falls off logarithmically; in *E* intervals are as follows in msec.: Make (*M*) to *1*, 5.1; *1–2*, 4.5; *2–3*, 24.6; *3–4*, 24.0; *4–5*, 21.8; *5–6*, 30.3; *6–7*, 38.5; *7–8*, 27.5; *8–9*, 36.3. Time for all records can be gauged from this.

significance relative to the conditions determining slow repetition are (1) that it is subject to the same control by anodal polarization as is fast repetition and (2) that the mechanisms producing slow and fast repetition probably are not very different, since, as in the present case, the two types are in evidence in different fibers of the same preparation.

But what is it that determines when the repetition in response to the action of a rectangular constant current is going to be slow and when fast? Though we have no definite answer to this question, we nevertheless would like to consider it for a moment. Such repetition as is determined by recovery of a fiber through the usual supernormality must of necessity be fast, since the maximum of excitability then is attained in from 5 to 6 msec. There are, however, types of supernormality with a period that is slower than 200 per sec. For example, there is apparently a "second supernormal phase" (Gasser, 1935*b*) which can attain its maximum in about 0.1 sec. If this second supernormal phase, or some other recovery process equally slow, could set the pace for repetition a rate of 10 per sec. would result, which is almost as slow as physiological repetition ever gets to be. But it is clear that should there be a first supernormal phase also, the second, if it is to set the pace, would have to rise higher than the first. At present, however, our information relative to these supernormal periods is insufficient to supply a basis for a discussion of their rôle, if they play one, in repetition. It remains for future investigation to determine definitely the mechanism responsible for the repetitive stimulation of nerve fibers that occurs in the normal discharge of their functions.

Just one word more on this subject. It has been stated that one of the rôles of anodal polarization in converting the usual single response of a nerve to a rectangular cathodal current into a repetitive response probably consists in causing recovery from a response to reach normal through a supernormal period. We recall this in order to indicate its possible relation to a phenomenon recently described by Matthews (1933). Matthews was recording, in the cat, the discharge along a muscle-sense fiber that was induced by stretching a muscle. The relatively slow rate of physiological activation obtained. Then the muscle was asphyxiated. After an interval the discharge suddenly changed to the rapid variety, one very different from the physiological type, the rate rising to 400 per sec. Now one condition that is known to convert recovery without supernormality into recovery through supernormality is a high concentration of hydrogen ions. Asphyxia, as is well known, increases the acidity of tissue. Here then is a sequence which could account for the pictures observed by Matthews: supernormal recovery presumably steps in when the tissue becomes acid, and sets a pace that is faster than that determined by the normal sensory mechanism, with the result that the repetition becomes rapid.

Manifestations of Segmentation of the Medullary Sheath

We turn next to the consideration of a phenomenon which was happened upon quite unexpectedly during the course of some experiments that had as their initial object the determination of the effect on propagation rate of altering the configuration of the action potential; but instead of gaining information relative to that problem some interesting pictures were encountered which possibly have a significant relation to the mechanism of the propagation of the nerve impulse (Erlanger and Blair, 1934).

To change the configuration of the action potential the nerve, the phalangeal preparation, was polarized continuously at the lead, sometimes cathodally, sometimes anodally, and note was taken of the effect these procedures had on the configuration of the conducted spike of the most excitable axon. The principle of the electrical circuit that was employed is illustrated in the upper diagram of Fig. 56, where can be seen the stimulating coil connected with one end of the nerve, and the recording instrument and polarizing circuit connected with the other. The lower diagram shows one of the circuits as actually used.

When the proximal lead, *P*, was polarized anodally and the action potential was monophasic, that is to say, when the distal electrode was on killed nerve as shown in the

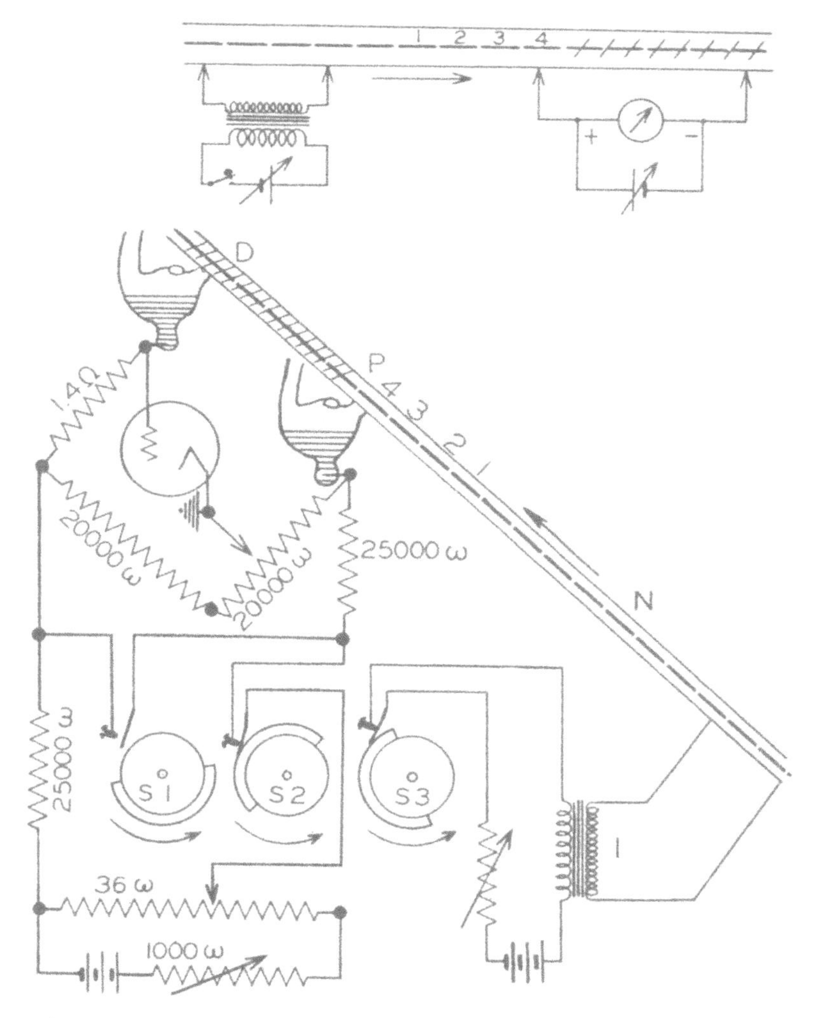

Fig. 56. Above: Diagram of a medullated axon in a nerve, illustrating in simple terms the stimulating, polarizing, and lead circuits employed in eliciting the nodal phenomenon.

Below: The same in detail.

diagrams, a curious series of pictures was obtained as the polarizing potential was increased. One series is shown in Fig. 57. Here the tracing of the normal axon potential is shown as *A*. With increasing polarization this spike first increased in height and became longer in duration, as in *B*. These are the well-known, the usual, effects of anodal polarization. The unexpected was the development of a notch (labeled *4*) on the ascending limb. This notch mounted faster relatively than the spike increased in height and became later and later until suddenly, and at a critical polarization strength, the part of the record above the notch dropped out leaving below it the lower, but characteristically shaped action potential completed by the dotted curve. By holding the polarizing current at this critical strength the spontaneous changes in the fiber's excitability now caused the picture (*B*) to vacillate between the large spike with its notch and the lower one without a clear notch. Now, upon resuming the increase in polarization, the process just described repeated itself but on a lower scale, as in *C;* again at a critical polarization level, higher, of course, than the first one, the part of the record above the notch, *3*, broke away leaving behind a still lower, but still typically shaped, axon potential, the one completed by the dotted curve. And this performance repeated itself

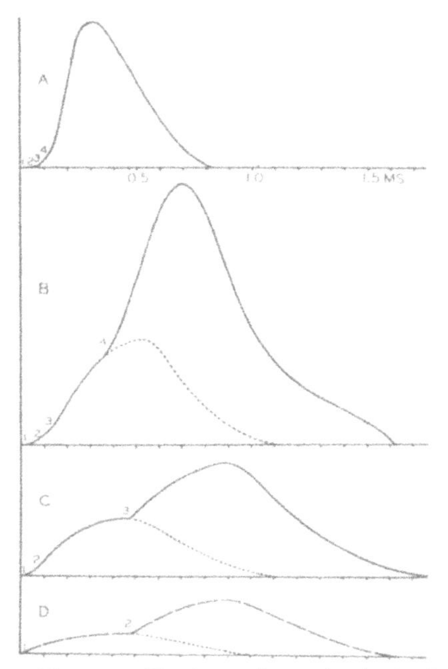

FIG. 57. Tracings of records obtained from the axon diagrammed in Fig. 56, as anodal polarization increases.

still another time with further increase in polarization, as seen in *D*. By this time, though, what was left of the action potential was so low that it could scarcely be distinguished from the noise level of the amplifier. If there were any further repetitions of the phenomenon they could not be recognized. Actual records of a process similar to that shown as line tracings of records in Fig. 57 are reproduced in Fig. 58.

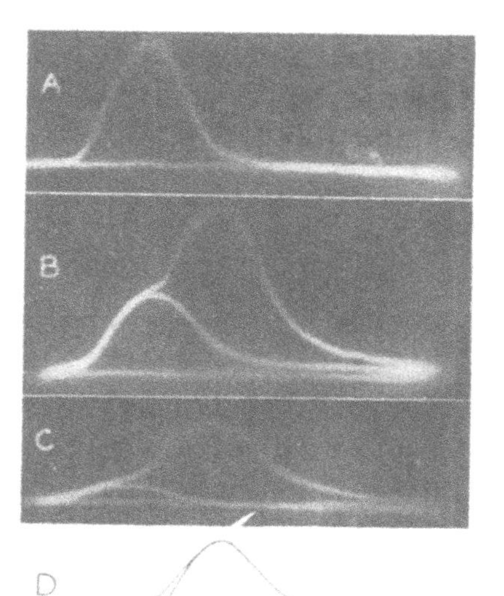

FIG. 58. Records from an axon as anodal polarization is increased at the lead. The time is given in *D*, and also a subtraction illustrating the method of obtaining the configuration of a segment spike, but one anodally polarized.

The clue to what is happening in such experiments is given by the results that are obtained when both of the leads are on intact nerve and the record consequently is diphasic. When, under such circumstances, the anode is at the lead that is proximal with respect to the stimulated point, as was the case for the record of Fig. 59, the negative or upward-directed part of the diphasic potential changes as the polarizing current is increased just as it does when, under the same circumstances, the lead is monophasic; that is to say, this part of the record becomes higher, develops the notch, and the upper element suddenly drops out

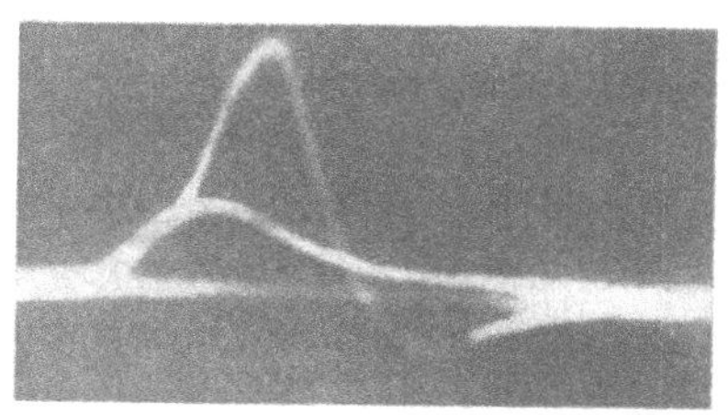

FIG. 59. Showing that a diphasic spike of an axon becomes monophasic with the first default induced by anodal polarization at the proximal lead.

leaving a lower spike. But now, at the same instant, the positive element, the downward-directed phase, also disappears; the record becomes completely monophasic.

The results obtained with the polarizing current reversed, i.e., with the anode at the distal lead, also are of interest from the standpoint of interpretation. Now, as polariza-

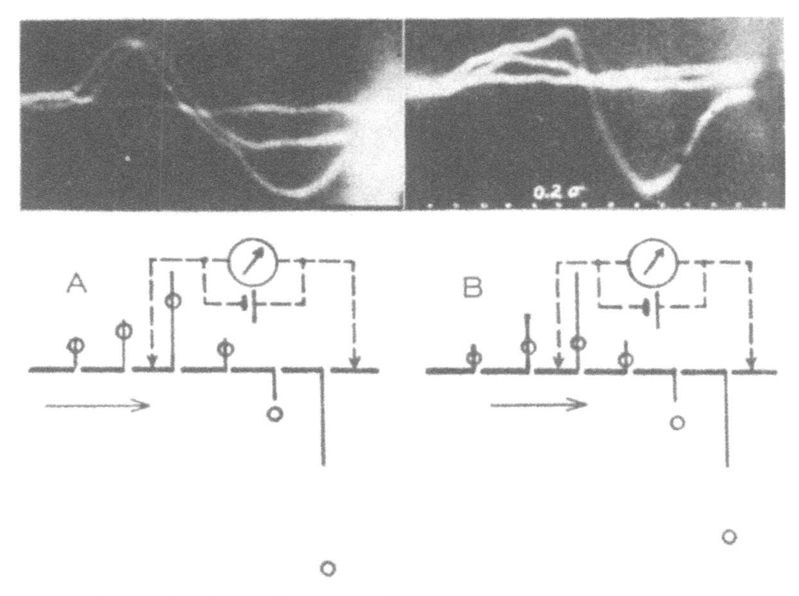

FIG. 60. Left: When the axon spike is diphasic, anodal polarization of the distal lead (see *A*) induces in the positive phase three successive defaults comparable with those seen at the proximal lead in Fig. 58. Only the first default is shown here.
Right: Under the same circumstances when the defaults begin to affect the negative (upward) phase the positive phase is completely eliminated.

tion is gradually increased, the positive phase, the downward-directed phase, increases in amplitude, develops a notch (as in the left-hand record of Fig. 60) and then the part beyond the notch suddenly disappears. With further increase in polarization still another but less ample piece of the positivity disappears in similar fashion; and the last trace of positivity disappears, again, at the instant the last and highest negative element drops out, as may be seen in the right-hand record.

Now how are we to account for these results of anodal polarization? Undoubtedly, through the establishment of localized blocks. The blocking unquestionably is comparable to that occurring in the classical experiment of Pflüger. In that experiment a nerve-muscle preparation is used, of which the muscle serves the purpose of the electron oscillograph as detector of the nerve impulse; and in that experiment the cathode of the polarizing current takes the place of our stimulating inductorium. With the make of the constant current the impulse initiated at the cathode travels toward the muscle, but is blocked at the intervening anode by the depression induced there by the strong anodal polarization. So in the experiments just described, the successive defections that occur as anodal polarization increases must be attributed to blocks developing in succession along the axon, and we have next to find some way of accounting for the seriatim blocks.

The responses we are recording are from medullated axons, the sheath of which is arranged on the fiber in segments; and the first thought that suggests itself is that the phenomena we have been describing are manifestations of this segmentation of the medullary sheath. As a matter of fact, all of the results can be made to fit the simple hypothesis that anodal polarization blocks as a result of effects exerted through the nodes of Ranvier, and that the effective potential associated with the excitation wave traveling along the axis cylinder leaks out to the leads on the nerve as the wave passed nodes of Ranvier. In other words it would appear that owing to the high resistivity of myelin, the effective passage of electricity in either direction between the axis cylinder and the exterior is via the nodes of Ranvier.

Now how does this hypothesis fit the facts? Segment length in the larger medullated fibers, the kind we are concerned with here, averages about 1 mm. The interlead distance usually has been such that it would subtend 5 or 6 or 7 such segments. The conduction rate in the fibers we are dealing with usually is of the order of 20 m.p.s. The usual conduction time per node, therefore, may be put down as 0.05 msec. The potential developed by the active locus traveling along the fiber, as indicated in Fig. 61, leaks out of nodes to produce a potential difference between the leads *P* and *D* (proximal and distal); the magnitude and the direction of its successive effects are indicated diagrammatically by the lines erected at the nodes. Potential also leaks out of nodes reached by the impulse earlier than

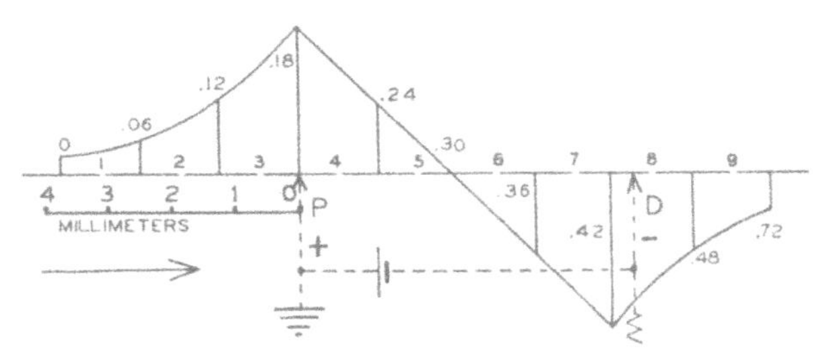

FIG. 61. Diagram indicating how nodes might account for the results obtained through anodal polarization of an axon at the proximal lead.

node *1*, but it is only from *1* that the potential spreading down the nerve affects *P* appreciably; it then produces a potential difference between *P* and *D* of the height indicated by the vertical; and it is that potential which records at that time. Later, by 0.05 msec., the active locus reaches *2* and begins to affect *P* through node *2*. We do not know definitely how much greater the effect exerted on the leads through *2* will be than through *1*, but, as we shall see, a doubling of the potential per node (probably about 1 mm.) just about fits the results, and the diagram has been constructed accordingly. The potential effective at the leads therefore doubles again when the active locus reaches node *3* and again when, 0.15 msec. after having passed node *1*, it reaches node *4*. The effective amplitudes and the sign of the potentials that become effective through the nodes that lie between the leads, *P* and *D*, probably are correctly indicated by the graph; and they would develop at the times indicated. The potential elements from all of these nodes must combine to form the effectively smooth diphasic potential that is recorded under normal circumstances.

Now how will polarization with anode at *P* affect the picture? Node *4* then will be the one most directly exposed to the anodal effect, and other nodes will be affected less in proportion to some function of their remoteness from *P*. Consequently *conduction* will be retarded most at *4*, and component *4* also will be the one to show the greatest anodal increase in amplitude. When the anodal polarization first reaches blocking intensity it will block at *4*, the impulse can no longer pass beyond and, of the entire action potential, the contributions from nodes *1*, *2* and *3* alone will remain. Consequently the highest element beyond the notch drops out and all of the later elements also; and the action potential becomes monophasic. But at this polarization strength the process from node *3*, due to current spread, is beginning to be appreciably affected. Conduction there is slowing and potential amplitude is increasing, and this again is intensified by further anodal polarization until element *3* is blocked out; and by the same token nodal elements *2* and *1* also eventually are eliminated. All of the records we have shown can be accounted for in this manner.

Thus far, for the sake of simplicity, the nodal contributions to the action potential have been represented as straight lines. Actually the action potential at each node requires time to attain its maximum and to decline. The temporal configuration of a nodal contribution, but polarized to be sure, and therefore longer in duration than a normal nodal contribution, can be determined by direct experiment. Given a monophasic record to which, say, nodes *1*, *2*, *3*, and *4* are contributing, polarization to the intensity at which component *4* drops out will leave a record consisting of components *1*, *2*, and *3*. In Fig. 62 we have in the dotted curves the tracings of a record changing in this manner from the whole, to the lower one. What has dropped out to leave the lower dotted curve is the contribution of *one* node. And if we subtract from the whole what is left after the defection, the remainder must have

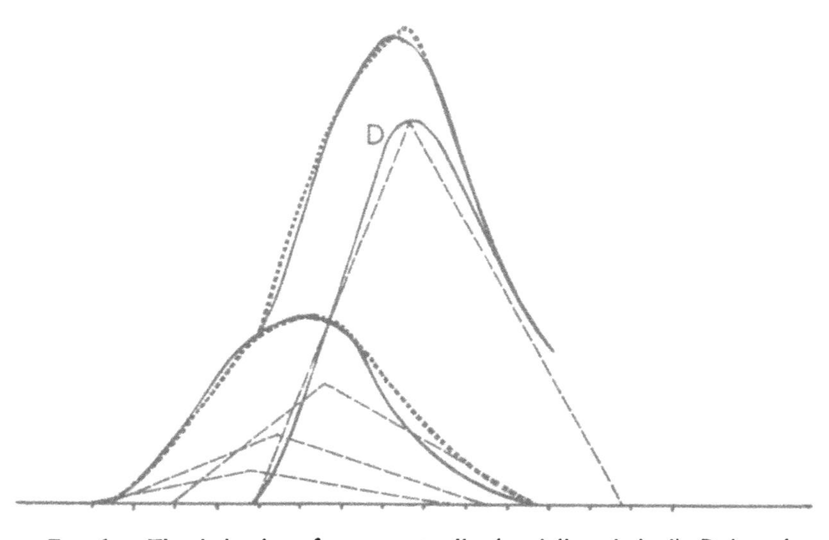

Fig. 62. The derivation of a segment spike (anodally polarized), *D*, by subtracting from the entire action potential the spike potential left after the elimination of the higher segment spike. (See also Fig. 58.)

the configuration of the potential from a single node. The result is *D*. This it will not be necessary to describe. But it may be worth while showing how the idea this procedure suggests can be used to derive the duration of a nodal contribution to a normal axon potential. To this end, a record was made of a monophasic axon spike to which, there were reasons for believing, five nodes contributed as illustrated by the lower diagram of Fig. 63; the record itself is shown as the broken curve. The determined conduction rate was such as to indicate that the internodal time intervals were of the order of 0.05 msec. Five triangles, spaced temporally accordingly, that is by 0.05 msec., and with altitudes proportional to the respective hypothetical nodal potentials, were then given by trial and error, such rising and falling times (the same for all) as would make their sum match the record. The triangles that accomplished this best, producing the solid line curve, had crest times

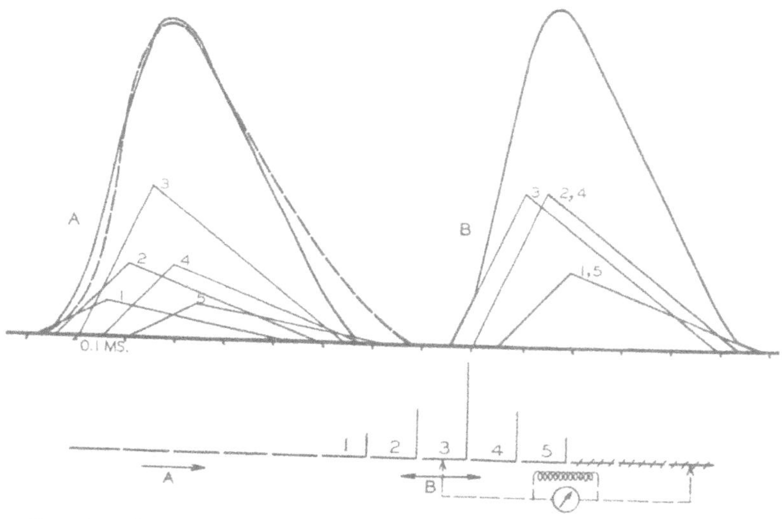

Fig. 63. Diagrams illustrating a method of deriving the nodal components of the axon spike: *A*, when it reaches lead *B* from *A*; and *B*, when it is initiated at the lead, *B*. Five nodes are contributing.

of 0.15 msec. This value is regarded as the nearest approach yet attained to the time to maximum of local activity in an axon.

We have now considered in some detail one of the possible ways, perhaps the simplest, of accounting for the experimental data of the present topic. There are, however, still other ways out which we may now proceed to discuss. I am sure you are acquainted with the Lillie iron-wire model of nerve propagation, a device which has been developed to illustrate one of the hypotheses of impulse transmission in nerve. An iron wire, you will recall, is exposed to the action of strong nitric acid which coats it with a layer of iron oxide. Thus coated, the wire is not attacked when placed in dilute nitric acid. But now exposing the iron at a point by removing the oxide coat there, permits the acid to react locally with the iron. As a result that point becomes positive electrically with respect to outlying points and a current flows from the bared wire to the immediately adjacent coated wire, reduces the oxide and exposes the wire there, while recoating and inactivating the bared spot. The wire newly bared then becomes the seat of a positive potential. And so chemical reaction and associated electrical current initiated by the scratch are propagated steadily along the model.

Now Lillie (1925) found that when the coated and immersed wire is threaded through segments of glass tubing the propagation of the reaction no longer proceeds steadily along the wire, but in jumps from intersegment space to intersegment space so that propagation becomes very much faster than along the wire when unenclosed. And it would appear that all of the results we have depicted in nerve are entirely compatible also with progression of the excited state in jumps from node to node.

But if progression in nerve is from node to node, and in the manner just described, it would have to be accomplished through restimulation by eddy currents flowing from node to node outside of the segments, just as happens in the segmented iron-wire model. In other words the process that determines impulse propagation in a fiber would have to operate through structures that are foreign to the fiber. From the standpoint of teleology it is hard to believe that this is the case. It seems much more reasonable to suppose that a nerve fiber conducts by means of a self-contained mechanism; that it contains within itself everything that is necessary for the performance of its own proper function. There is, indeed, evidence indicating that outside eddy currents generated by fiber activity are not strong enough to be of physiological significance; and it may be worth while devoting a moment to the experiment that had as its purpose the investigation of this question (Blair and Erlanger, 1932). The phrenic nerve is formed in the main by two trunks from the cervical plexus. By stimulating one of these trunks one can send an action potential along about half of the fibers composing the nerve and can ascertain whether the excitability of neighboring, but inactive, fibers is raised by such eddy currents as may be originating in the active fibers and flowing through the neighboring tissue and fibers. But it has been found that the inactive fibers do *not* respond any more readily to an induction shock while an action potential is passing alongside of them in other fibers than at other times. This and similar observations strongly suggest that currents eddying around nodes are of no significance physiologically.

If, then, progression of the nerve impulse is from segment to segment and is not effected by a transfer of energy around the segments, the transfer that leads to restimulation must take place through them, and across their ends. When one contemplates nerve from this standpoint, one is constrained to regard as the function of the medullary sheath not the direction of currents to the outside, but rather the prevention of leakage of current from the axon, with its consequent dissipation of energy and risk of stray effects. As a matter of fact segments give the impression of being insulated, isopotential cells, with constricted ends consisting possibly of semipermeable partitions, in contact at the nodes. Is it not likely that a localized chemical change or other potential source in such a cell would practically immediately raise the potential uniformly throughout the cell's entire mass, so that the speed of potential rise throughout the segment would approximate the speed of the chemical reaction or other process initiated locally? Then, if restimulation could occur across the nodal interfaces, the resulting picture of the electrical propagation would be just the one we have recorded.

It is seen, therefore, that our observations can be made to fit all three of the conceivable restimulation mechanisms of impulse propagation, namely, (1) continuous restimulation and saltatory restimulation of two kinds, (2) from node to node by currents eddying around the segments, and (3) from segment to segment across the nodes. Such relevant evidence as we have been able to acquire all favors the view that progression is saltatory, though it does not help to decide by which one of the two possible mechanisms it

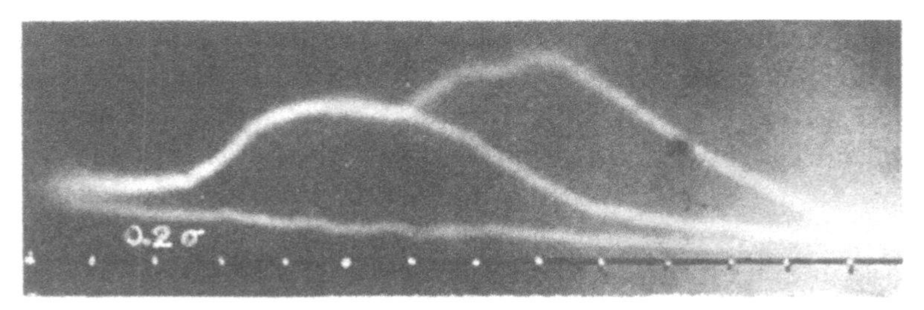

FIG. 64. A segment spike starting about 2 msec. later than the crest of the potential preceding it. The dot intervals are 0.2 msec.

proceeds. The evidence is as follows. If progression is saltatory, segment length should have but a slight effect on the interval between the unit waves seen in the records of polarized spikes; whereas if progression is continuous the intervals should be proportional to the lengths of the segments. Judging by Hatai's measurements (1910), variations in segment length are frequent enough and wide enough to be recognizable in our records if progression were uniform. Yet only once in the course of a great many observations have we seen any considerable departure from uniformity in wave spacing in anodally polarized fibers.

That progression is by restimulation at or across nodes is indicated by the observation that anodal block never develops until the start of the wave about to be blocked has fallen back as far, at least, as the crest of the preceding segmental response, and often not until there is an interval between crest and start, an interval that has been as long as 0.2 msec., as may be seen in Fig. 64. Under but one set of circumstances could a continuously progressing point source of potential fail to produce a continuous *rise* of potential at the lead as it approached it, namely, if the duration of the rising limb were shorter than the conduction time in a segment. Practically, that seems to be an impossible state of affairs. And so it is felt that the weight of evidence favors saltatory progression of the nerve impulse in medullated fibers and by a mechanism that causes the segments to act as units.

IV

SEQUENCE OF POTENTIAL CHANGES

NERVES transport to the central nervous system quantitative information about states of the parts of the body, and in the reverse direction, they mediate finely graded activity of the effector organs. Complex and varied events are brought into being in the terminal apparatus, but for the inauguration of all this complexity of effect there serves a very simple mechanism of transmission. What this mechanism is, it is unnecessary to relate to an audience familiar with the beautiful experiments of Professor Bronk and Professor Adrian, which have shown that variation in the messages carried is completely accounted for by the number of impulses per fiber and the number of fibers active. Discrete signals arranged in ever changing spatial and temporal patterns effectively convey all the information which nerves need carry in the fulfillment of their function in the coördination of the organism.

In the terminology used in connection with the physiology of axons these unit signals are called spikes—or spike processes, if we wish to differentiate the event from the electrical sign of the event. Spikes as events may be compared to the ticks of a clock. Both are but signs of activity in an underlying mechanism, and both are ultimately dependent upon stored energy for their maintenance: the one upon the energy of a coiled spring or lifted weight, the other upon the energy derived from oxidative metabolism.

It follows then that if spikes are but manifestations of activity in the inherent mechanism of nerve fibers, the story of nerve is by no means told when the spikes have been described. We need to know something about the mechanism which produces them—how it is maintained, its capacity for work, and when and how the work is paid for.

The examination of nervous activity with the foregoing questions in mind has brought out ample evidence that the nervous mechanism involves processes other than the one immediately concerned with impulse transmission. In fact, activity can best be regarded as made up of a chain of events of which the spike process is but a part. The evidence for the several processes is found in the many studies which show that associated with activity in nervous tissue there occur:

An increase of metabolism and an accumulation of identifiable catabolites,
An increase of heat production,
An alteration of excitability,
An alteration of electrical impedance, and
A sequence of electrical potential changes.

Out of the interpretation of the studies along these lines and the analysis and correlation of the information which they contain must be evolved our notions of what may be the nature of the processes. Therein lies a great difficulty. All the evidence is indirect and the validity of the deductions drawn from them uncertain. Consequently our ideas of how the nervous mechanism operates will probably be in a state of evolution for a long time to come.

Strictly speaking, the body of our knowledge about nerve consists of the sets of data which have been obtained by the various methods devised for the purpose of measuring the manifestations of nerve activity; but insight into current thought about nerve functioning cannot be gained by confining our attention to sets of data alone. The experimenter is forced to set up in his mind models of nerve which integrate the facts and give them meaning: and interpretations are made largely in terms of these model nerves—the only "nerves" which are understandable. If all the facts were known, the model nerve would be as complex as is the actual nerve and identical with it. In reality models of this kind are the crudest sort of simplification. Their merit lies in the questions about nerve which they suggest. Experimental data are the answers of real nerve to these questions.

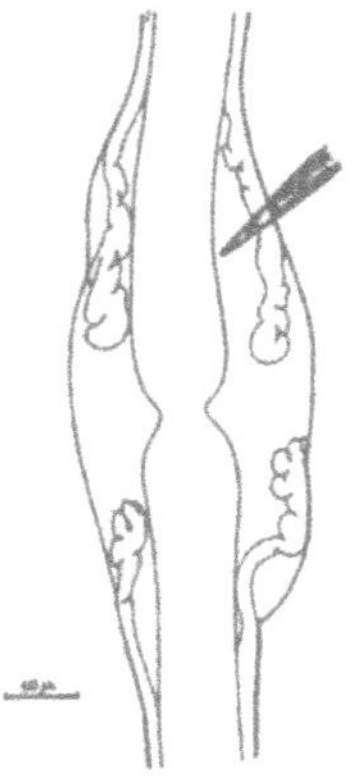

FIG. 65. A frog medullated nerve fiber into which saline has been injected through a micropipette. The neurilemma, the myelin, and the axon are separated by the fluid. Although freed of its sheaths, the axon has maintained its form. (de Rényi, 1929.)

The part of the nerve fiber responsible for conduction is the axon. Structurally the axon is a thin, cylindrical strand of protoplasm having in microdissections of freshly isolated frog fibers, according to de Rényi, the consistency of a soft jelly (Fig. 65), but so fluid in the large nerve fibers of invertebrates as to be extruded whenever the surface of the fiber is traumatized (Bozler; Young, Fig. 66). Fibrils are present in the axial part of the cytoplasm of invertebrate fibers, but are not found in living frog fibers (de Rényi) and hence cannot be regarded as being the conducting element. A more probable nutritive function has been assigned to them by Parker.

Nerve conduction must, therefore, be thought of as taking place in a tissue of fluid or semifluid consistency, and whatever of structure is present must be inherent in the colloidal properties of its substance, or in its surface film. Like the surface of all living cells, the surface of a nerve fiber has the power of selecting the ions which can pass its border. Hence differences of concentration arise between the inside and the outside of the fiber, and a resistance is imposed to the passage of an electric current. When a portion of the surface is destroyed and electrodes are placed on the fiber—one on the injured portion and one on an intact part—a difference of potential is seen to exist between the two points, the injured region being negative to the uninjured part. The seat of the potential is at the uninjured surface, the injured portion serving only to afford an electrical

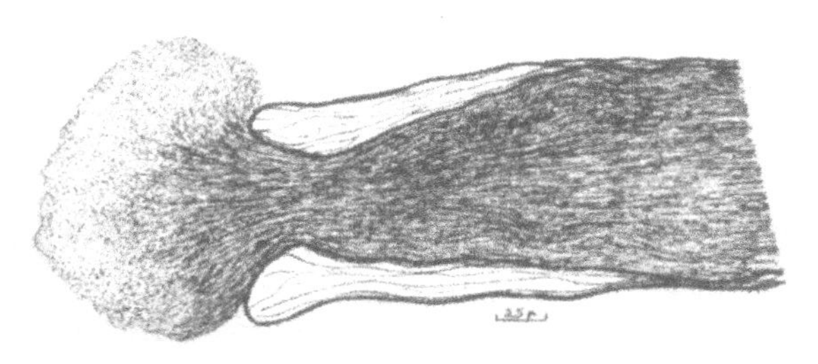

FIG. 66. Giant fiber from the stellar nerve of Sepia. Twelve minutes previously the fiber had been cut across, and the contents are flowing out from the end of the connective tissue sheath. (J. Z. Young, 1934.)

connection to the inside of the fiber. Displayed in physicochemical terms, the arrangement (Fig. 67) is one in which electrodes dip into electrolytes of two concentrations separated by an intervening partition, one concentration, C_1, representing the electrolytes inside the fiber and the other concentration, C_2, those outside.

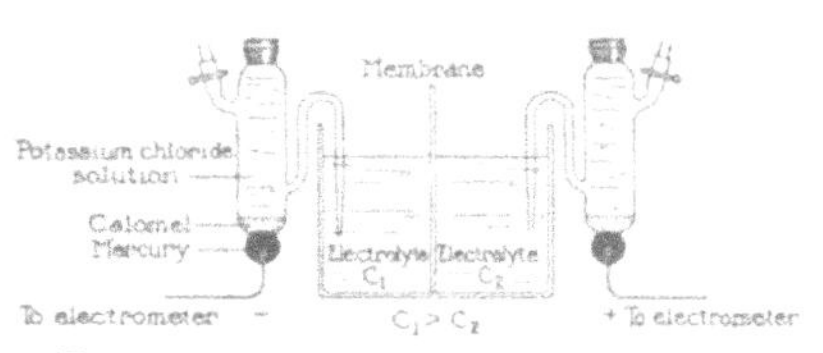

FIG. 67. Concentration cell, with electrodes in an arrangement resembling the chain occurring when leads are taken from the intact surface and the killed end of a nerve fiber.

A physical chemist desiring to reveal the greatest potential difference which the two concentrations of the electrolyte could possibly produce would place between them, as the intervening partition, a sheet of metal corresponding to the cation which the solutions have in common. His potentiometer would then show a voltage determined by the expression:

$$E = 58 \log \frac{C_1}{C_2}$$

where E is the potential in millivolts and 58 is a constant holding for room temperature and univalent ions. In other words, if one concentration were ten times the other, the potential would be 58 millivolts.

Calculations from data obtained by Fenn and his colleagues indicate that the normal composition of nerves is as stated in the following table:

TABLE III

Probable Electrolyte Distribution in Frog Nerve
Concentration in millimols per cent *

	Inside Axons [1]	*Outside Axons* [2]
Potassium	17.6	0.27
Sodium	3.7	10.8
Calcium	0.96	0.21
Magnesium	2.4	0.31

* Calculated from Fenn, Cobb, Hegnauer, and Marsh (Table 6).

[1, 2] Called respectively by the authors the potassium space and the sodium-chloride space.

The potassium concentration inside the fibers is, therefore, 65 times that which it is outside, and the ratio is sufficient to permit a potential of 118 millivolts.

Maximal values of the demarcation potential, as measured between an intact portion of a nerve and a freshly injured portion, are approximately 30 millivolts. On account of the shunting circuits existing within the tissue, the actual potential drop across the membrane must be larger than the measured value; but, even when allowance is made for shunting, a sufficient margin remains to permit the assumption that the difference of concentration is great enough to account for the potential found. In some way or other the surface film is able to serve as a means of revealing a large fraction of the theoretically maximum potential.

In the past a most troublesome problem has been, how a potential of this kind could be made to appear in a tissue in which a metallic partition is impossible. That difficulty has now been set aside by two ingenious models, one invented by Beutner and the other by Michaelis. In one model, the potential is produced by employing oils of special chemical composition; these oils, in the presence of electrolytes, assume properties resembling the metal corresponding to the electrolyte. In the other model, the potential is revealed by interposing a specially prepared collodion membrane having pores so fine that it acts as an ionic sieve. Because of the charge on its surface, it is permeable only to cations, as is nerve; and it gives, therefore, a diffusion potential under the condition that one of the ions is immobile, a condition which permits a diffusion potential to have the same magnitude as a concentration potential.

Physicochemical chains of this kind give us confidence that we can build a nerve model with a nonmetallic surface. It is unimportant which of the two systems is selected, for, as Michaelis has stated, in a surface layer that has at most the thickness of a few molecules, the difference between the two systems would largely disappear. All arguments become couched in molecular terms, and the properties of the surface must depend upon its molecular structure and on the forces existing about the molecules.

For mechanical reasons it would be easier to construct a nerve model, as Ebbecke has done, by molding a Michaelis membrane in the form of a tube. A model of this type, filled with electrolytes as they occur in the axon and surrounded by electrolytes as they occur in the tissue spaces (Fig. 68), would give a potential of the right sign and magnitude if an electrode on the surface B were connected through a

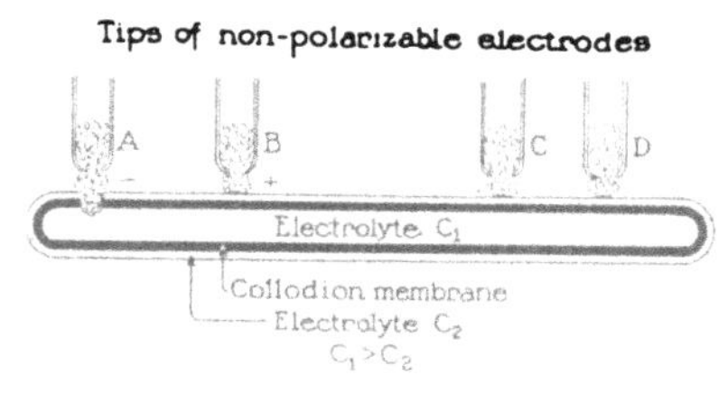

FIG. 68. Model having some of the properties of a nerve fiber.

potentiometer to another electrode penetrating to the interior *A* (the opening in the membrane corresponding to an injured portion of a nerve fiber). A somewhat crude imitation of the spike of the action potential could be produced if a momentarily existing hole were made under the electrode *B*, for then the potential drop across the membrane under *B* would be destroyed and the difference of potential between the electrodes *B* and *A* would momentarily disappear. The electrotonic or polarization potential of nerve could also be imitated by employing a polarizing current to the model through the electrodes *C* and *D* (Labes and Zain, Ebbecke); but at this point the analogy with nerve would begin to break down.

No matter how thin we conceive the walls of the model to be, their immutable structure serves to differentiate sharply model from nerve. In order to make the theoretical model resemble nerve more closely, we should have to permit the composition of the surface to change, both with the composition of its interior and with its surroundings.

We have seen that the cytoplasm of nerve has a fluid consistency. Hence its molecules are free to move. According to the thermodynamic principle known as the Gibbs-Thompson rule, any substance in the interior of a liquid which will reduce the free energy of the surface of the liquid, will be concentrated in the surface. The composition of the surface is, therefore, determined by the composition of the fluid from which it is formed; and as the rule is one having universal application, it must hold also for the cytoplasm of nerve. We must think of the surface membrane, then, as a structure which is in equilibrium with the interior of the axon, or at least as one which deviates from equilibrium only because, for dynamic reasons, equilibrium cannot be attained.

The theoretical nerve now has become much more complex. The surface film is formed out of the interior of the fiber, and because of its molecular structure it has the power of detecting ionic concentration differences at its boundaries and reporting their magnitudes in terms of electrical potentials. In its normal state it reports a potential representing a large fraction of the theoretical maximal potential which could be produced by the difference in concentration that is present. But when the surface is disturbed, it reports a much smaller potential, just as would happen if a loose collodion membrane replaced a tight collodion membrane separating two solutions.

A surface film formed out of cytoplasm and in equilibrium with it is endowed with an important theoretical property. Any modification of the surface would upset the equilibrium, thereby bringing about a change in the underlying cytoplasm in adjustment to the alteration of the surface. For instances of change in cellular activity produced in this way we must turn to the pharmacological literature, for in pharmacology it is a well recognized principle that stimulation with all its train of metabolic events may be effected by pharmacodynamic agents that, under the conditions of the experiments in which they are tested, could hardly be expected to penetrate the surface of the cell. If pharmacodynamic agents at the cell surface can produce these deep-seated effects, it is quite reasonable to expect that a disturbance in the surface as great as goes with the spike potential can also produce them. Any alteration of the metabolism in the fiber would then bring about a condition such that, when the spike process is ended and the surface comes to be reformed, it would be reformed about a medium with a new molecular constitution and consequently have a composition differing from the one obtaining at the start. In the new form, its effectiveness in reporting differences of concentration in terms of electrical potential would be altered slightly; thus, even if no shift in the distribution of electrolytes had occurred during the period of spike production, an alteration in the potential reported would be present and would persist until the whole system returned to its original state. Viewed in this way, the potentials which follow the spike may be regarded as qualitative indicators of the course of chemical reactions within the nerve fiber, and as such they become quite intelligible.

We know that chemical changes are set up by nerve activity, because measurements of the oxygen consumed and carbon dioxide liberated have proven that an increased metabolic rate occurs, and because some of the more stable metabolites have been identified. From the thermal analyses made by A. V. Hill and his colleagues we also have information about the time at which the metabolism takes place. Except for a few per cent at the most, the heat is produced after the spike production is completed. At this time, however, the potential change has not yet ended. A slowly progressing potential of small magnitude still remains. While no experiments have been done in which the heat production and the potential change have been determined on the same nerve, the correspondence between the durations which have been reported for the two events is sufficiently close to lead to the belief that they are associated.

We shall now trace the course of all the potential changes which occur in a nerve following the application of a single electrical stimulus of short duration, as for instance a break induction shock.

The Polarization Potential

The first visible electrical event is commonly called the shock artifact. It is made up of two parts. One part, caused by coupling between the leads from the nerve and the stimulating apparatus, varies according to the nature of the apparatus used and has no physiological importance. The other part, caused by polarization of nerve structures, has considerable interest because of its relation to the mechanism of electrical excitation.

The polarization or, as it is often called, the electrotonic potential, is referable to the condenser-like properties of the nerve-fiber surface, and it may be imitated with any of the core models. (The tube made with a Michaelis membrane is a special case.) To make its origin clear, the membrane may be set forth in condenser form (Fig. 69) in a manner resembling the one originally proposed by Hermann. The capacity of the membrane is represented by condensers; and conductivity inside and outside of the fiber and across its surface is represented by resistances.

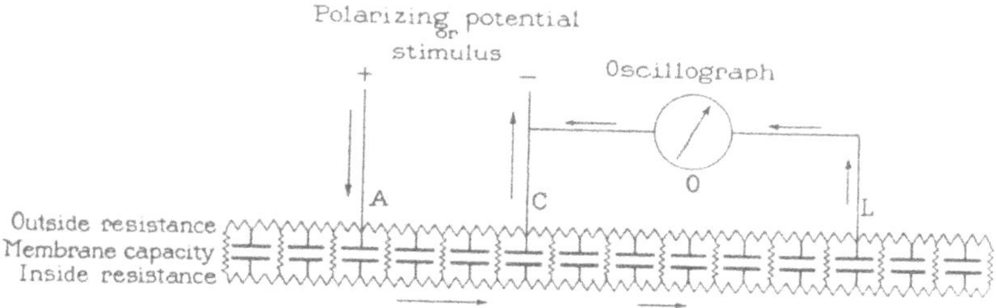

FIG. 69. Diagram representing the surface of a nerve fiber as a leaky condenser, in a manner resembling the schema of Hermann and others. Electrodes are drawn in connection with the model in positions fitting the arrangement which is used in nerve experiments when the stimulating cathode is also one of the leads. The arrows show the direction of current flow during the passage of the polarizing current. In the branch L O C the arrows also show the direction of the current after the polarizing circuit is interrupted.

Let us assume that a stimulating current is applied from a potential source through electrodes *A* and *C*, which are anode and cathode respectively, and then consider the potential changes which would take place at the cathode, since it is at the cathode that an action starts in nerve. At the first application of the potential, current would flow readily through condenser *A*, the internal resistance, and condenser *C;* and only to a small extent through the adjacent condensers. As a result, the potential difference between the terminals of condenser *C* would rise and interpose a counter electromotive force in the circuit, blocking flow through the condenser. The current would then increase in the adjacent condensers and in turn increase the potential difference between their terminals; thus the electrotonic potential would spread along the model as a pseudowave, just as Bogue and Rosenberg have shown that it spreads along nerve fibers.

A circuit equivalent to the one used in nerve experiments, when a lead from the stimulating cathode is desired, is drawn in Figure 69 and marked with the letters *L O C*. It will be noted that lead *C* is common to the exciting (stimulating) circuit and to the registering circuit containing the potentiometer *O*. When a current from a constant potential source is started in the exciting circuit, there is at first only a small potential difference between *L* and *C;* but as the surface of the model near the cathode becomes polarized (that is, as condenser *C* is filled), the potential difference rises rapidly and finally reaches a steady value. Interruption of the exciting current at this stage leaves the system with all the condensers charged; *C* is negative to *L*, and remains so until the potentials of the condensers are equalized through the high resistances. If the exciting current be of brief duration, the period during which the condensers are charged is short, while their discharge, as in the preceding case, has a duration determined by the circuit constants; thus the polarization potential resulting from an induction shock as recorded by the potentiometer *O* would have a rapid rise and a more gradual fall.

Some variation from the form of the polarization potential derived from the condenser model occurs in nerve, because of the fact that in a stricter analysis the nerve would have to be represented by a more complicated network (Bishop, 1929), and because the capacities in nerve are believed to be polarization capacities and consequently vary with the time; but the general form which the model predicts is followed (Fig. 70). The polarization potential appears in its pure form as long as the stimulating induction shock is below threshold. As soon as the shock becomes greater than threshold, the potential is complicated by the spike rising above it.

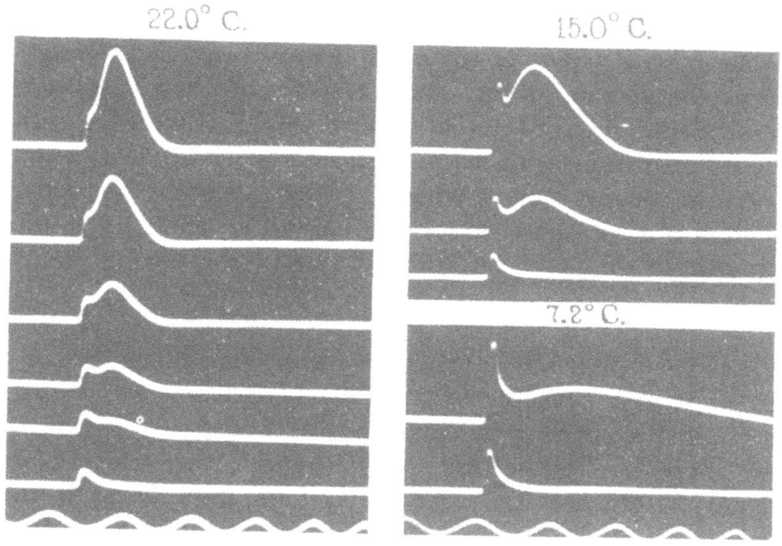

Fig. 70. Polarization and action potentials in a spinal root of the bullfrog when the stimulus and the lead are at the same point, as in Fig. 69, and when the stimulus is a break induction shock. Lead interpolar distance 13 mm.; stimulus interpolar distance 5 mm.; time, msec. Records at three temperatures. The lowermost records at each temperature are below threshold. The shock strength at 22° is given the value 1 and other strengths are reported in relation to it. At 22° C. the shock strengths in turn are 1, 1.16, 1.25, 1.4, 1.68, and 2.10. Threshold at 1.03 (record not shown). In the top record the response is maximal. At 15° C.; 1.4, 2.1, 4.2. Threshold at 1.52. At 7.2° C.; 2.8, 5.6. The record with strength 1 at this temperature was exactly as at 22° C. In addition to the points mentioned in the text these records show: Addition of fibers as the strength of shock is increased, shortening of the make-response time with increasing strength of stimulation, rise of threshold with cooling, and also a prolongation of the spike and an increase in area.

The only potential known to precede the beginning of the action is the electrotonic potential. As Professor Erlanger has explained, the application to a nerve of a shock or a rectangular current at a strength insufficient to excite, produces during and after the passage of the current a sequence of excitabilities the course of which is determined in part by the nerve and in part by the form of the current. Processes occur which cause the excitability to rise above normal during the flow of the current—and in the case of brief impulses for a short time thereafter—and then to sink below normal after the current has been broken. It might be expected that some characteristic of these processes would cause them to reveal themselves by a potential sign, but up to the present no potentials have been seen which have the proper time course. Either they do not exist, or they are so small that they cannot be detected in the presence of the large electrotonic potential.

The Spike of the Action Potential

It is very instructive to watch the potential changes in a nerve at the cathode of a stimulating current. If one starts with an induction shock considerably below threshold and increases its voltage progressively, the electrotonic potential which is produced increases parallel with the size of the shock until threshold is reached. Then a very small increase in the applied voltage causes a sudden jump in the potential, and the added potential has a form which is not foreshadowed in the slightest degree in the previous potential-picture (Figs. 70 and 71). One cannot watch the process without becoming convinced that the spike of the action potential is not a transition from the polarization potential. There are no transitional stages. It appears as though released by a trigger mechanism. After a certain amount of preparation, the spike potential bursts forth full-fledged. Furthermore, the electrotonic potential and the spike vary quite differently when subjected to temperature changes, the electrotonic potential, for instance, being almost unaffected when the nerve is cooled, while the spike is greatly prolonged (Figs. 70 and 71).

As Nernst originally proposed, the primary event after the application of a stimulating current must be a transport of ions. These ions accumulate at barriers and produce the electrotonic potential. At the same time, the accumulation initiates two processes which are not detectable electrically—one directed toward excitation, the other

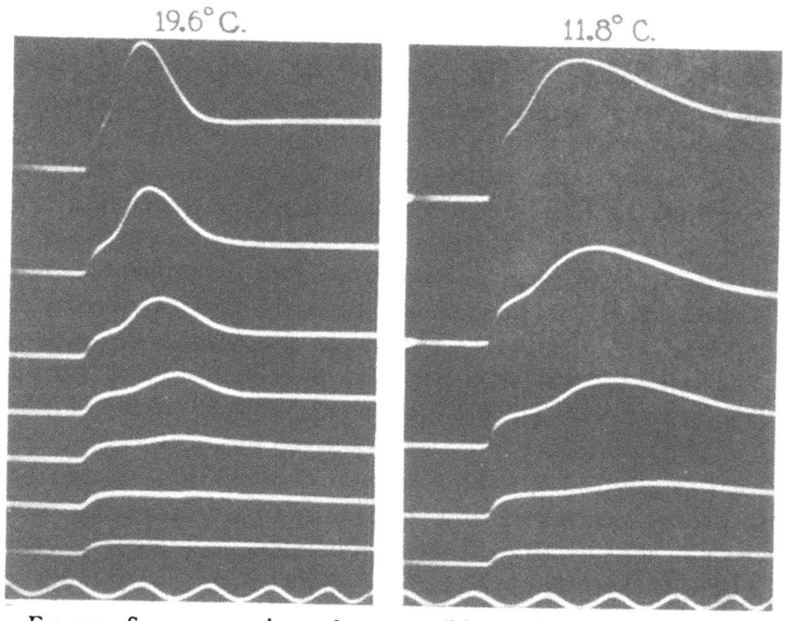

FIG. 71. Same preparation and same conditions as in Fig. 70, except that the stimulus is the make of a rectangular current. The lower records at both temperatures are subthreshold. They are produced by the same applied potential and are alike in form. Strengths of current at 19.6° C; 1, 1.1 (threshold), 1.33, 1.55, 1.89, 2.12, 4.45; at 11.8°: 1, 1.33, 2.00, 2.78, 4.45. The threshold at 11.8° is at 1.1. It may be noted that it is not higher than at warmer temperatures as holds for induction shocks (Fig. 70). The long utilization period as compared to induction shocks will be noted and also the larger dispersion of utilization periods among the fibers. See also Bishop (1928) and Schmitz and Schaefer (1933).

(called accommodation) opposing it. If the time course of the stimulating current is such as to permit the algebraic sum of the two processes to reach a value sufficiently in favor of excitation, an event occurs which brings about the appearance of the spike potential. As to what may be the intimate nature of the event, we have no notion; in fact, we cannot even state whether the immediate origin is chemical or physical.

Two observations bear on the question: the temperature coefficient of the spike duration, and the heat production; but neither one throws the balance of the argument to the one side or the other. The temperature coefficient of the rate of rise of the spike potential, $Q_{10} = 2$ to 3, so often cited as evidence that spikes are the result of a chemical change, is without crucial significance in view of the possibility that the limiting factor in determining the rate of rise may be freedom of movement of the molecules in a lipoid surface film. Lipoids have temperature coefficients of viscosity quite large enough to permit of the formulation of a theoretical interpretation completely physical in nature. The same lack of crucial significance also holds with respect to the small amount of heat which may be liberated during spike formation. In this instance A. V. Hill (1932), mindful of the condenser-like properties of the surfaces in the fiber, has pointed out that the heat evolved may as well be interpreted as representing the energy of a discharging condenser as that of an exothermic chemical reaction.

The outstanding quality of the spike process is the fidelity with which it maintains a constancy of its own characteristics through forces inherent in itself. All-or-nothing in nature, with its size dependent upon the state of the nerve and not on the form of the impulse current producing it, the spike process differs strikingly from the phenomena occurring subthreshold which follow the strength of the applied current very closely. The time course of the spike is maintained with particular tenacity. Conditions which modify the potentials following the spike, and which even change the magnitude of the spike itself, are without effect upon the duration. The spike thus stands at the center of the nerve's organization as the particular feature the integrity of which must be preserved, if the nerve is to maintain its specialized function of the transmission of messages. The other processes may be considered as existing for the purpose of maintaining the conditions necessary for spike production.

Excitation is usually thought of as being accompanied by depolarization of the plasma membrane, and the view is now well substantiated for nerve through Lullies' measurements of the nerve's resistance during activity. Conduction of a low frequency alternating current is facilitated during activity, as the theory of depolarization demands. In the plant *Nitella*, in which the slowness of the action permits a more detailed examination, Blinks has indeed been able to show that at the height of the action the polarizability (and hence the polarization) is lost. In nerve, however, the depolarization can hardly be considered to be complete, as is often supposed. If it were complete, it would have to follow the demarcation potential when the temperature is changed, whereas what happens is that the height of the spike changes rapidly with the temperature at readings below 20° C. (frog A fibers), (Fig. 72), while the demarcation potential changes much more slowly (Verzár, Bremer and Titeca). Thus, the colder the nerve, the smaller relatively is the amount of depolarization during the course of the spike in terms of the magnitude of the demarcation potential.

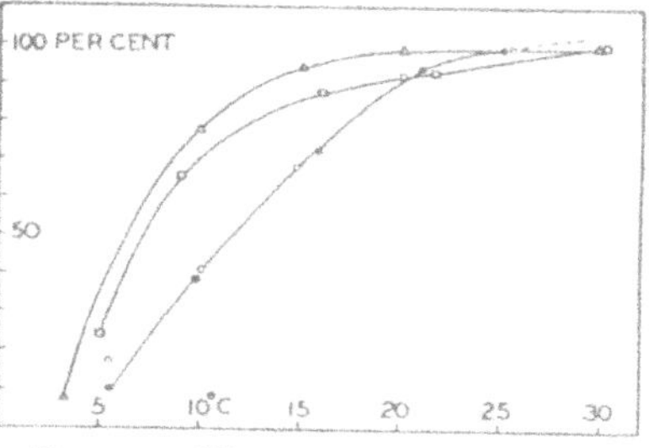

FIG. 72. The effect of temperature on the height of the spike of frog A fibers. The values are recorded in percentages of those observed at the highest temperature examined, about 30° C. □, △, June frogs; ●, ○, October frogs. Rapid changes with temperature such as these are characteristic of physical systems; e.g., viscosity of paraffins and fats, or electrical conductivity of tetraethylammonium bromide in a mixture of alcohol and lecithin. (Gasser, 1931.)

The fact that depolarization during the spike is not maximal can also be shown in other ways. If a single spike is followed by a second and a third one, instead of all three rising to a common ceiling, the second and third add to the negative after-potentials of the preceding responses. Therefore, the depolarization at the crests of the first two spikes at least cannot be maximal. Finally, Bishop (1932*a*) has shown that when a nerve is poisoned by certain drugs the spike falls off more rapidly than does the demarcation potential. In his experiments at the start of the blocking of conduction the demarcation potential was reduced 10 per cent and the spike potential 30 per cent.

The next question to be answered is: When does the spike end? Two difficulties stand in the way of satisfactorily measuring the duration. The end of the spike overlaps the beginning of the after-potentials; and the region of the combined potential with which we are dealing is the one that suffers the greatest distortion from the diphasic artifact of the spike. The question is best investigated with the three-electrode system and, as described by Bishop (1932*b*), with cocaine carefully placed about the part of the nerve in contact with the distal lead, in order to make the record of the potential as monophasic as possible. If a frog nerve be used and a shock employed which brings into action a fraction of the A fibers, it is found that the spike rises to its crest in 0.3 msec., and then falls again to within a few per cent of zero potential at 0.9 msec. At this point the rate of restoration is markedly retarded, so that the potential curve thereafter approaches the base line very slowly. As it now becomes fused with the negative after-potential, the location of its ending cannot be made out; but some idea of how far the potential goes may be gained by making use of the fact that the spike is constant in form, while the negative after-potential is variable. Subjecting the after-potential to experimental variations makes possible a series of action potentials made up of a constant and a variable, from which the constant portion carried from one action potential to the other can be evaluated. When worked out in this way, the tail of the spike of frog A fibers is found to be traceable for about twenty-five crest times. At this point it may be considered as having ended, or as undergoing a transition to positivity, depending upon the interpretation of the further course of the potential picture.

The mode of ending which has just been described is known to hold only for frog A fibers. Mammalian A fibers have not been examined in this way, and C fibers in both forms follow a quite different course. No reliable figures are as yet available for the duration of the C spike. Potentials from single C fibers have never been seen. Leading from the cathode is disturbed by the large artifact connected with the high threshold of stimulation, and even a short distance of conduction introduces a large amount of temporal dispersion, the latter at 5 mm. of conduction being theoretically as great as at 300 mm. in A fibers. In my experience, the best preparation for the study of frog C fibers has proven to be the splanchnic nerve of the bullfrog, because of its small content of fibers of other sorts. Leads made from small strands at high amplification, so as to approximate the conditions of single fiber recording, give rising phases of about 3 msec., and a similar figure has been obtained in the few cases where successful leads from the cathode have been possible. With the ordinary method of recording and 5 mm. of conduction, rising phases of 10 to 15 msec. are obtained. Records from the splenic nerve of the cat made under similar conditions, but at 37° C., show a rising phase of 2 to 3 msec. or longer. In single units, however, the rising phase probably does not last much over 1.5 msec.

As first shown by Bishop (1934), C fiber spike potentials end by passing abruptly through zero. They are then succeeded by a well-marked positive potential (Fig. 77A, 1). Positivity begins after two or three recorded crest times have elapsed and proceeds rapidly to a maximum, the latter occurring at about 50 msec. in frog fibers and at 7 to 25 msec. in mammalian fibers (5 mm. conduction). Following the maximum, the potential returns slowly to zero along a decremental curve. Whether the sequence of events is a spike overlapping and summing indistinguishably with a positive potential produced by another process, or whether the whole may be considered as a continuous process, cannot be decided at present.

The Negative After-Potential

Overlapping and immediately following the spike there occurs a potential which shows considerable independence of it—the negative after-potential to which allusion has been made previously. The latter is extremely variable in form. In freshly mounted frog sciatic nerves it may last for only 20 to 30 msec. (Fig. 73); but as experimentation proceeds it grows in size and duration, and still greater augmentation may be brought about by poisoning the nerve with drugs, so that the potential may reach ten per cent or more of the value attained by the spike, and the duration may be extended to a matter of minutes.

Fig. 73. Relation of the spike to the negative after-potential in A fibers of a fresh preparation of the sciatic nerve of the frog. Data from Figure 2, Gasser and Erlanger (1930). In this condition the after-potential is small. After experimentation it becomes higher and longer, but the spike does not change.

If a series of records be obtained of the variations of the action potential of a single preparation—preferably of one undergoing the changes induced by the alkaloid, veratrine—it is found that the spikes of all the records are superimposable, but that near the end of the spike there appears a point at which the curves diverge. When the curves are drawn one upon the other, the after-potentials of the graph appear to pivot about the point at which they become differentiated from the spike (Fig. 74).

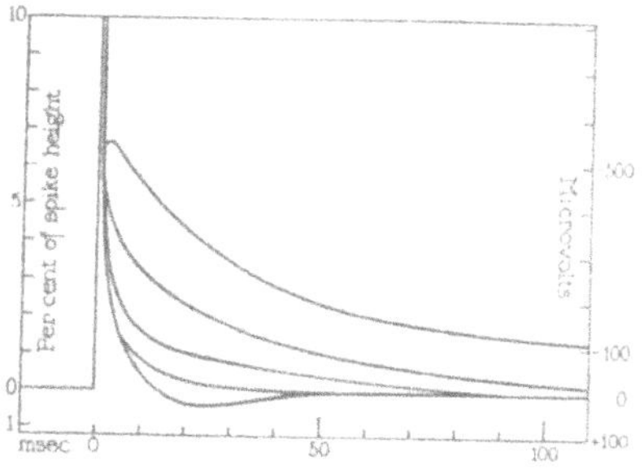

FIG. 74. Successive stages in the development of the negative after-potential in a veratrinized phrenic nerve, presented as superimposed drawings of records, 37° C. The spike leaves the base-line and goes above the figure. It then returns to the figure and is continued by the negative after-potential. In the unpoisoned nerve the latter is short and is succeeded by a positive phase (lowermost line). After addition of veratrine the successive lines show how the negative after-potential develops. The topmost line is from a record made soon after a tetanus. (Gasser and Grundfest.)

In most preparations spike and after-potential fuse so perfectly that the juncture is entirely imperceptible, and were it not for exceptions in occasional instances and for developments appearing after special procedures, the relation between the two might have been misinterpreted. Occasionally frog sciatic preparations yield evidence which, although not striking, is unmistakable in its indication that the negative after-potential starts out with a rising phase of its own (Fig. 75); and responses set up in both mammalian and amphibian nerves, which have been treated with veratrine and tetanized during the period immediately preceding the observation, show the development to a marked degree (Fig. 76).

Fresh preparations of frog C fibers develop no potential which can definitely be identified as the negative after-potential, unless a slight suggestion of a convexity in the trough of the positive potential may be so interpreted (Fig. 77A, 1). The capacity for production of a potential of this kind is present to an extraordinary degree, however.

After treatment with veratrine, the negative after-potential rises out of the trough of the positive potential with a rising phase conspicuously setting it off from the spike; and it has the appearance of being added to and concurrent with the potential from which it rises. As the poisoning proceeds, the negative after-potential grows progressively,

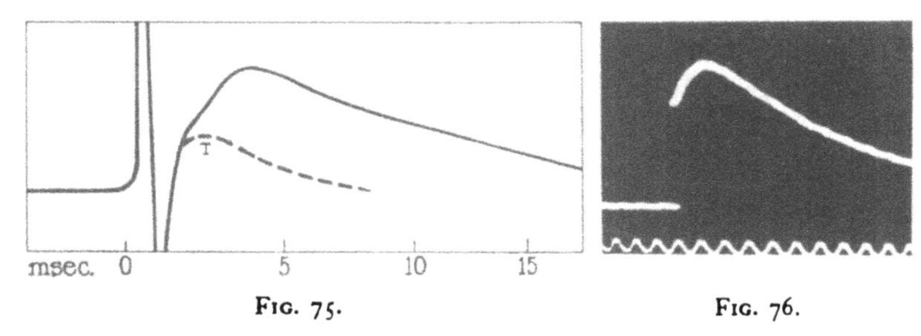

FIG. 75. FIG. 76.

FIG. 75. A portion of an action potential from the sciatic nerve of the frog, showing a rising phase in the negative after-potential. The drawing starts with the spike, 1.4 per cent of the height of which is visible; and as the spike is not strictly monophasic it is succeeded by a diphasic artifact. Following the diphasic artifact the potential becomes negative again, and above the point, T, a discontinuity appears which marks the juncture of the spike and the negative after-potential and identifies the end of the spike. The diphasic artifact has cut into the tail of the spike separating off a remnant labeled T, in analogy with the T wave of the electrocardiogram. From this point onward the spike potential must decline and its probable course is indicated by the broken line. When the end of the spike is subtracted from the combined potential, the negative after-potential still has a rising phase. The original record was taken from a normal nerve, with 3 mm. of conduction and a shock strength bringing out 60 per cent of the A spike. (Gasser and Graham, 1932.)

FIG. 76. Record of a single response of a veratrinized nerve, set up soon after a tetanus. The potential change starts with the spike at the break in the line, but it is not visible until the beginning of the negative after-potential. The latter starts with a well-defined rising phase. Time 60 cycles.

and it may finally attain a value which is higher than the corresponding spike (as recorded with the dispersion obtaining at 5 mm. of conduction) (Figs. 77 B and C). Preparations of mammalian C fibers, as for example the splenic nerve of the cat, behave in this regard in quite the same way.

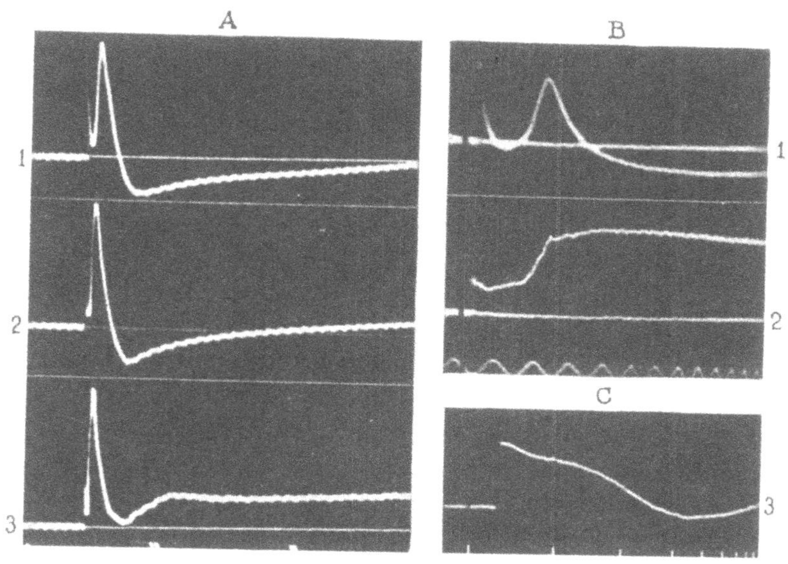

FIG. 77. After-potentials in C fibers in the splanchnic nerve of the bullfrog. In all the records the first elevation, which is only slightly visible, is caused by fibers faster than C, and the sharp elevation is the C spike. A 1, 2, and 3. Normal response (1), and two stages of veratrine poisoning. In A 2 the long, slow rise of the negative after-potential will be noted, actual negativity being reached only at the end of the record. Beyond the record, the potential will become positive again. A 3. In this record the rising phase of the negative after-potential is well marked. Time 0.2 sec. The ripple is due to a. c. B 1, 2. Normal C spike and the same in later veratrine poisoning with an after-potential larger than itself. The rise of the base-line preceding the C spike is the after-potential in the faster fibers. Time, 40 msec. C 3 shows the continuation of B 2. Time, 10 sec.

The Positive After-Potential

In our sequence of potentials we now have the spike and the negative after-potential. Next in order comes the positive after-potential. The latter in single responses of frog A fibers is extremely minute, not over 10 μv for the whole A group, and it may be absent altogether. In this respect, frog A fibers stand in striking contrast to frog C fibers, in which the large positive potential following the spike must be considered as a positive after-potential.

Mammalian A fibers present some differences from frog A fibers. The positive after-potential appears with great regularity at a fairly constant, instead of a very variable, interval after the start of the spike; and it has a much larger size and definite duration. Connected in all probability with these characteristics of the positive after-potential there are also differences in the appearance of the negative after-potential. The latter appears to be complete in 10 to 20 msec. at the beginning of the positive potential; but, inasmuch as in some preparations there are signs of its reappearance at the end of the period of positivity, the time of its apparent ending must merely indicate a point at which the processes tending to produce the potentials with the two signs come into balance. Positive and negative after-potentials can hardly be considered separately, because there is every indication that in A fibers, as was previously noted in C fibers, the processes for which they stand run concurrently. It will also be seen later that there is some interrelation between them.

Under conditions resembling closely a nerve's normal physiological environment the positivity attains a maximum at approximately 30 msec., and then decreases until restoration reaches completion at approximately 70 msec.; but in conditions other than normal, particularly if the hydrogen ion concentration in the medium about the nerve be changed, the values are somewhat different (Lehmann).

Potentials Associated with a Tetanus

Impulses are not carried singly by nerves functioning in the body, as they are in laboratory experiments. On the contrary, especially in sensory fibers, spikes course over the nerves in rapid succession. Adrian, Cattell, and Hoagland show the impulses as following one another at the rate of 300 per second in a skin nerve of the frog, and Matthews has described rates of 450 per second in nerves from asphyxiated mammalian stretch-afferent receptors. According to Bronk and Stella even the relatively small interoceptive nerve fibers from the carotid sinus may have occasion to carry 170 impulses per second. The state of the after-potentials during and after a tetanus is, therefore, of much greater physiological importance than the one following single responses. Such states may be examined by subjecting a nerve to a train of shocks from a thyratron stimulator and measuring the after-potentials after amplification with a direct current amplifier.

The behavior of the negative after-potential as it is seen in frog A fibers holds in general also for mammalian fibers. With the aid of the diagram in Fig. 78 and the records of

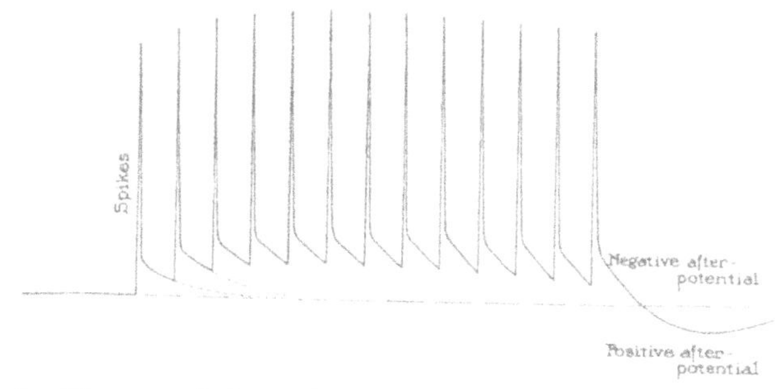

FIG. 78. Diagram of the action potential during a tetanus. For obvious reasons the after-potentials are drawn disproportionately large with respect to the spikes.

Fig. 79 it will be possible to follow the steps by which the changes in the after-potentials take place. The first spike appears with its negative after-potential, and in the course of the latter the second spike starts. As the second spike rises to its full height above the existing potential level, it is added to the residual negativity of the after-potential left from the first response; and the absolute negativity of the second crest is, therefore, higher than that of the first one. A slight increase in the absolute level of the maximum of the second after-potential usually occurs also, but there is no increase in duration. As a consequence, the area of the after-potential from two responses is much smaller than the sum of the areas belonging to the individual responses (Amberson and Downing). The incompleted portion of the first response is in effect omitted. In other words, the negative after-potential does not accumulate in proportion to the number of spikes. As the tet-

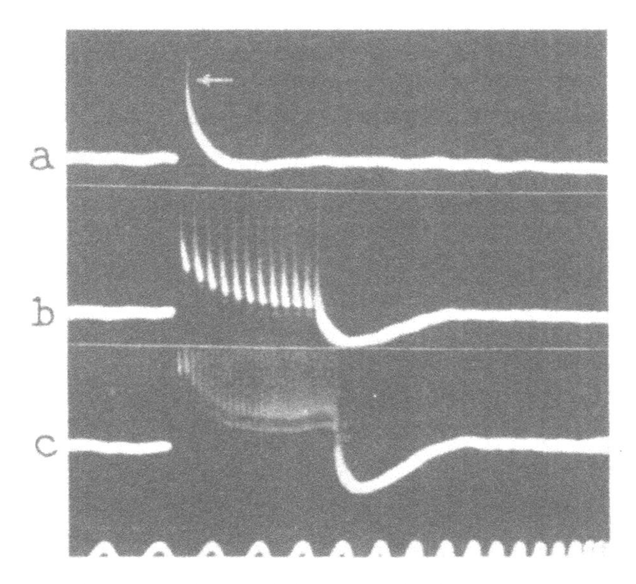

FIG. 79. After-potentials in a single action of a fresh phrenic nerve at 37° C. (*a*), and the alterations produced by short tetani at 180 and 350 per second (*b* and *c*). Time, 20 msec. The records have been made strictly monophasic by application of cocaine to the distal lead. If there had been a diphasic artifact it would have appeared as a sharp incisure at the position indicated by the arrow, which is the junction of the spike and the negative after-potential.

anus proceeds, however, some progressive increase in the heights of the after-potential maxima occurs, giving the appearance of staircasing. Parallel with the increase, the spike crests also rise. Then both crest-heights and after-potential maxima reach a ceiling. In some nerves the ceiling is maintained; in others a progressive fall occurs, the difference between the two cases being determined by the balance between the existing tendencies to produce negative and positive after-potentials.

Until a maximum is reached, the magnitude of the positive potential following a tetanus increases with the frequency and duration of the tetanus (Figs. 80 and 81). Even during the course of a tetanus, evidence of an augmenting tendency of the nerve to produce positive potential is apparent in that successive negative after-potentials appear to decline more rapidly (Figs. 78 and 79). At constant intervals of stimulation, as the positive potential

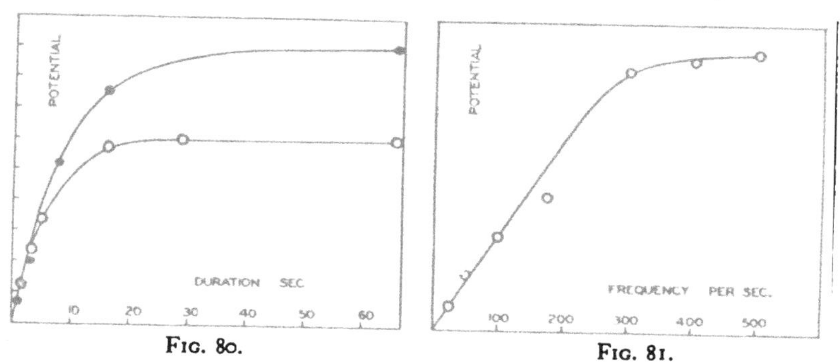

Fig. 80. Fig. 81.

Figs. 80 and 81. Size of the positive after-potential following tetani of different durations (Fig. 80), and of constant duration but varied frequency (Fig. 81). Frog sciatic nerves. Increasing the intensity of tetanization beyond the point which will produce a maximal positivity results in an extension of the duration. Gerard (1929) reports durations of 10–15 minutes in frog nerve. Gasser and Grundfest have found durations of 5 minutes in isolated mammalian nerve at 37° C.

process gains ascendency over the negative, the potential level following each constituent response falls to progressively lower values before the intervention of succeeding shocks. Reflecting this effect, the negativity at the spike-crests also undergoes a decline. The positions of the negative after-potential maxima become lower; and if the tetanus is stopped at any point, it is seen that the period during which the after-potential remains negative is much shorter than it is after an isolated response. (Compare the first and the last responses in Fig. 78; also *a* and *c* in Fig. 79.)

While the positive potential is characterized by a rapid increase to maximum followed by a slow return to resting potential, closer examination of it shows that it cannot be interpreted simply by the development and relaxation of a single process. Two parts appear with varying distinctness in frog nerve and with great distinctness in mammalian nerve.

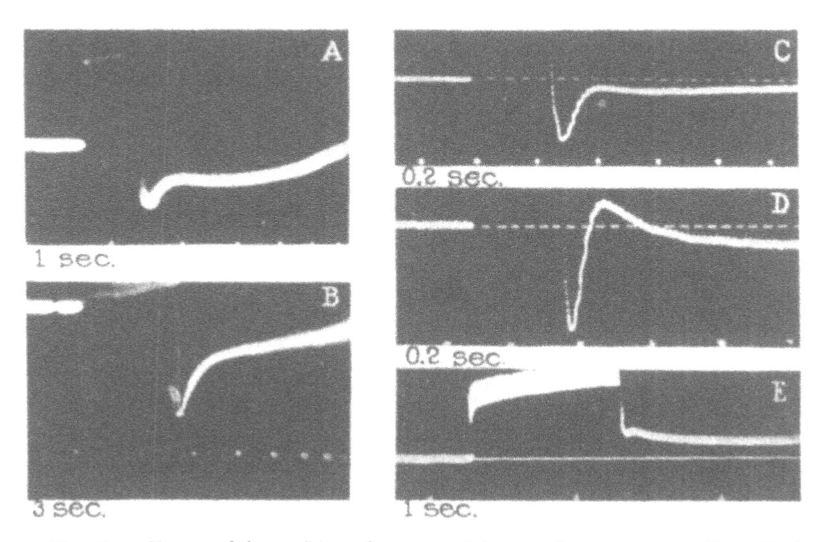

FIG. 82. Forms of the positive after-potential seen after a tetanus. Records A and B, frog nerve; C, D and E, mammalian nerve. In all records, the tetanus starts at the break in the line, and during the tetanus either nothing is seen or the bottoms of the curtailed negative after-potentials are visible. At the end of the tetanus the oscillograph spot drops sharply to outline the first positive notch, then after a period of rapid restoration marks out the second part of the positive potential.

No greater differentiation of the two parts may be present in frog nerves than is indicated by a transition from one velocity of restitution to another, that is, from a rapid velocity holding in the early period of restoration to a slow velocity in the later period (Fig. 82 B). In an occasional instance, however, the early period may be separated from the later by a maximum in the potential curve (Fig. 82 A).

In mammalian nerves, on the other hand, a maximum between the two parts is regularly present, and variation occurs only in that the absolute potential of the maximum shifts in accord with the amount of negative after-potential which may have developed. With a moderate development the crest is on the positive side of zero; with a larger development the crest is negative, but is succeeded by further appearance of positivity; while if the negative after-potential is very large, the whole of the first positive notch is negative to zero, and a swing of the potential to positive is delayed (Fig. 82 C, D, E).

Information derived from studies of the excitability of nerves *in situ* affords good reason for the belief that of the foregoing cases the only one which depicts normal nerve functioning is that in which the crest is on the positive side of zero. The others are outside of what may be considered to be the physiological range of the nerve; but they nevertheless serve a useful purpose in that attention is called to the composition of the normal cases by the exaggerated features of the abnormal ones.

Let us follow the development of the potential with an attempt at an interpretation of what is taking place. Tetanization increases the depth of the positive potential, which corresponds to the one seen in a single action, from an original value of 25 μv up to a value as high as 0.7 mv, the amount varying with the severity of the tetanus. Thus the first positive notch is produced. At the same time, the negative after-potential is increased, and depending upon the extent of the increase, the position of the notch is located at various levels. The variations observed make it clear that the action potential throughout this period is determined by two competing influences: one producing negativity and the other the notch. When the negativity is unduly developed, the action potential becomes negative for the second time. This observation affords a clue to the significance of the maximum between the two parts of the positive potential when the absolute value of the maximum is positive. The maximum is occasioned by the negative after-potential process.

The second part of the positive potential, if present in a single response of a normal nerve, is too small to be detected; but it, or one analogous to it, develops in a single response of veratrinized nerve following the period of augmented negative after-potential. When brought out by a tetanus, unlike the first part of the potential, the second part varies in length as well as depth in relation to the duration, and especially in relation to the frequency of the stimulation. While there is no valid reason for not entertaining the idea that the second positive potential is a direct result of the spike production in a tetanus, a more plausible interpretation seems to be that it is a consequence of the increased negative after-potential produced, just as is the positive potential, in a veratrinized nerve. If the latter view be correct, a statement of the whole sequence would be: Increased spike production increases both the first positive after-potential and the negative after-potential, and the process producing the latter in its later stages also produces the second positive potential. It, of course, would be understood that the transition from one to the other starts immediately and that continuation of the positivity results not only from something that the negative after-potential process leaves behind, but also from the continuation of the process itself.

The nature of this sequence of potentials may be somewhat clarified by placing it more explicitly in analogy with the sequence of potentials in a single action in C fibers in an early stage of veratrinization. It will be recalled that in the trough of the large positive after-potential, normally present in C fibers, a negative crest appears (Fig. 77 A2 and Fig. 95) and that with the appearance of this crest the total duration of the positivity is extended. The sequence of the potentials is then: spike, positive potential, negative potential, positive potential, and restoration to normal. Similarly the potentials in tetanized A fibers take the course: spikes, positive potential—in this case made large by the tetanus—intercurrent negative potential—also resultant from the tetanus but comparable to the veratrine-produced potential in C fibers—and finally the terminal positivity.

Correlations

It was stated at the outset that our knowledge of nerve must be gained from a correlation of the manifestations of its activity. As the action potential is the one manifestation that defines precisely the time at which the events having potential signs occur, it logically becomes the manifestation most suitable for common reference.

Of the correlations made, the one which has been most extensively studied, the correlation with excitability, will be left for discussion in the next lecture and mention will be made here only of the relationship of the action potentials to heat production, to metabolism, and to the state of the demarcation potential.

Potentials, Heat Production, and Metabolism

Heat production, which is the only other sign of activity that is at all well localized as to time, has already been mentioned as paralleling the after-potentials. If on the basis of this parallelism a correlation is to be attempted between the two signs, the first postulate which must be made is that an exothermic reaction accompanies the second positive potential; otherwise the necessary potential duration would not be found. The question then follows: What is the relationship of the negative after-potential to heat production? At this juncture an interesting observation made by A. V. Hill (1933) may be recalled. After making his analysis of the time course of the recovery heat, he came to the conclusion that the process of recovery occurs in two phases, an early rapid portion measured in seconds, and a later slow portion measured in minutes. In the light of our present knowledge of the after-potentials, this division is not surprising, for the early course of the after-potentials certainly differs from the later course. The possibility at once becomes apparent that the early rapid portion may be attributable to the negative after-potential process, which on the evidence of the form of the action potential is operative at this time to a much greater extent than it is during the later part of the heat production; and that the later slow portion corresponds more exactly to the second part of the positive potential, deriving its existence either from the production of the latter or from a persistent, but attenuated and masked negative after-potential.

Although it is impossible to differentiate the negative and the positive after-potentials with respect to their chemical relationships, it is possible to state quite definitely that the after-potentials as a whole represent a process involving an oxidative metabolism. The evidence for this statement was obtained in some experiments conducted in 1933 by F. O. Schmitt and myself. The plan of the experiments was simple. It was argued that if a given number of spikes in a veratrinized nerve occasioned a greater metabolism than a similar number of spikes of the same size in normal nerve, the difference must be attributable to the after-potential; and the proposition was put to the test. In these experiments it was found that two spikes per second was a very favorable frequency for the tetanus, inasmuch as in unpoisoned nerve this amount of stimulation did not call forth any increase in metabolism which could be measured, while in veratrinized nerve it produced a very definite increase (Fig. 83). During the tetanus, the rate of oxygen consumption rose 10 to 15 per cent—as much as an unpoisoned nerve shows when the tetanus frequency is 100 per second—and the high rate continued for a time into the resting period.

At the time of these experiments the difference between the two conditions was thought to be only the amount of negative after-potential produced, and accordingly the increase of the metabolic rate was associated with the latter. That interpretation, however, can no longer be considered as completely valid. We now know that even during a tetanus the negative after-potential is accompanied by an increasing tendency to produce positive potential, and that after a tetanus the negative potential finally gives way to a positive one. There thus stands at the present time a complicating factor, which prevents the increased heat production from being associated exclusively with the negative after-potential. In view, however, of the possibility that the positive potential is resultant from the process causing the negative potential, the high metabolism of veratrinized nerves subjected to a tetanus seems to be traceable ultimately to the production of negative after-potential.

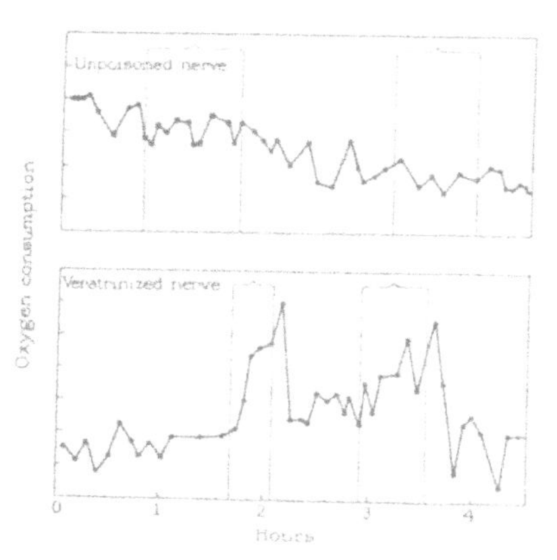

FIG. 83. Comparison of the increase in metabolic rate, produced by a tetanus of two shocks per second, in an unpoisoned frog's sciatic nerve with that in a veratrinized nerve under similar conditions. The nerves were tetanized during the intervals bracketed. The ordinates represent rate of respiration in scale divisions per minute. (Redrawn from Schmitt and Gasser, 1933.)

The association of the after-potentials with an increase in oxidative metabolism brings up the question of what may be the significance of the increase. Does it represent the energy of a mechanism actively intervening for a definite purpose in the chain of events which make up nervous activity, or is it simply a consequence of an oxidation of catabolites left behind after the spikes have been produced?

As far as the negative after-potential is concerned—and this is the only one that has been studied from this standpoint—there is good reason to believe that the process best fits the first possibility. This statement, however, is made in the face of a widely held opposing theory. Confusion has arisen because a very simple theory has been in conflict with a very stubborn fact. Let us consider first the theory and then the fact.

Placing a nerve in nitrogen causes the potential drop between the outside and the inside of the fibers to decrease, and a state is soon reached in which the nerve is no longer capable of transmitting impulses. Returning it to oxygen brings about a return of the normal polarization and a restoration of function—all of which means that a continuous supply of oxidative energy is necessary in order to maintain the integrity of the fibers. The theory is in line with this observation. It supposes that catabolites formed during the action are responsible for the negative potential, and that the potential persists until the catabolites are oxidized away.

Experimental support for the theory is found in the work of Levin and of Furusawa. The latter, by tetanizing crustacean nerve in nitrogen, produced a large negative potential ("retention"), outlasting the tetanus. Unless oxygen was readmitted to the nerve, no recovery took place.

The fact referred to was discovered by Amberson, Parpart, and Sanders. When a frog nerve is asphyxiated, the after-potential does not get larger because of accumulated catabolites, as the theory demands, but on the contrary undergoes a depression differentially greater than that of the spike. We have been able to confirm this observation repeatedly and in several ways. The relationship has been brought out most impressively in an experiment proposed by F. O. Schmitt. Nerve asphyxiated in the dark in a gas mixture made up of 96 per cent of carbon monoxide and 4 per cent of oxygen has its negative after-potential differentially depressed in the manner usual for asphyxia. Restoration is then effected by exposing the nerve to a bright light (Fig. 84). Both spike and after-potential increase, but the latter increases much more rapidly than the former, so that it far overshoots its original size. (Augmented negative after-potential is also characteristic of nerves to which atmospheric oxygen is restored after

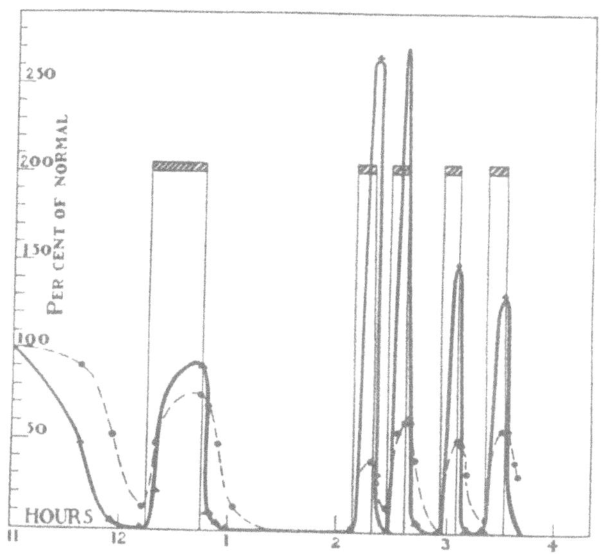

FIG. 84. Effect on the spike height and the area of the negative after-potential of asphyxiation by carbon monoxide in the dark; and the temporary restorative effect of brilliant illumination. The spike heights are shown by dots, and the after-potential areas by triangles. Periods of illumination are marked by rectangles with barred tops. (Schmitt and Gasser, 1933.)

a period of deprivation.) Through the alternation of light and dark periods, the cycle may be repeated a number of times. In every light period, in which the oxidative catalyst is freed from its poisoning by carbon monoxide, the negative after-potential appears.

The observations made on crustacean and on frog nerve thus stand apparently in opposition. Derived from the one is the view that the negative after-potential is a measure of the extent to which a nerve has not been restored; derived from the other, the view that the negative after-potential is the sign of a special process which may be connected with restoration. The derivations at least are not reconcilable. Only one conclusion seems possible, namely that the "retention" in crab nerve is not analogous to the negative after-potential of frog nerve; but the way in which the phenomena differ has not been delimited.

If in the interpretation of the phenomena taking place in frog nerve we follow the lead furnished by frog nerve and avoid inferences drawn from crustacean nerve—even though they fit in most alluringly with the membrane hypothesis—we can hardly escape the conclusion that active oxidation during the event itself is a necessary condition for the production of the negative after-potential. And when we consider this conclusion in connection with the observations made on the negative after-potential, which show that the latter does not start out at a maximum but undergoes a phase of development, that it is capable of independent variation, and that it accelerates the rate of recovery—we become increasingly confirmed in our conviction that the negative after-potential represents a separate event in the economy of the nerve.

The after-potentials are usually considered as being an expression of the process of recovery. This view is quite tenable, but only if it be recognized that the negative after-potential process accelerates recovery at the cost of an extra expenditure of energy, and that this expenditure adds to the total expenditure ultimately necessary. Nerves working in the body call on the mechanism to a limited degree only. Under the adverse conditions of isolation the mechanism comes into greater play, and in the exaggerated condition existing in veratrine poisoning the mechanism becomes very wasteful of energy and serves no useful purpose.

Alteration of the Negative After-potential

The negative after-potential may be either increased or decreased by experimental procedures. What happens to the positive potential at the same time is for the most part unknown. (By an increase or decrease is meant the variation relative to the change in the height of the spike.) In single responses of frog A fibers, ions of the alkali metals, hydroxyl ions, narcosis produced by aliphatic narcotics, and cooling depress the potential; while ions of the alkaline earths, hydrogen ions, warming, and notably veratrine increase it. H. T. Graham, whose articles have been drawn upon extensively for the statements made in the preceding sentence, has also found that the alkaloid, yohimbine, markedly increases the positive potential.

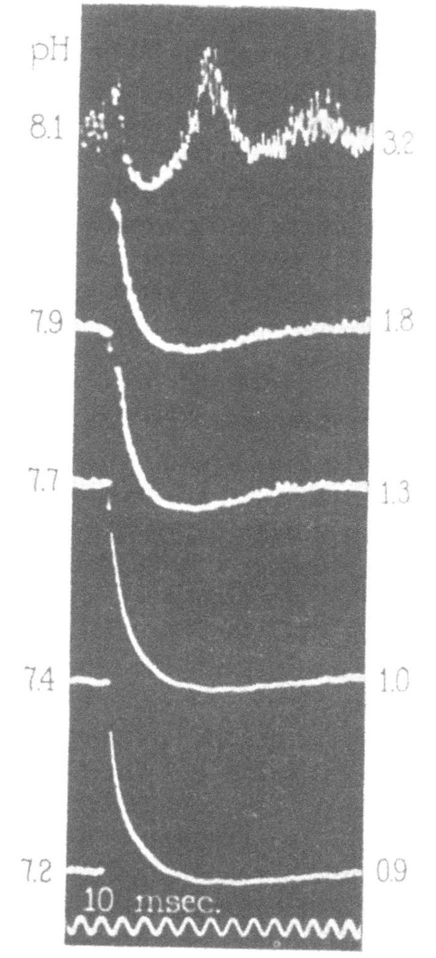

FIG. 85. Effect of changes in the concentration of hydrogen ions upon the after-potentials of mammalian nerve—phrenic nerve of the cat. The nerves were placed in a buffered Ringer's solution, and the reaction was changed by altering the mixture of O_2 and CO_2 with which it was equilibrated. The pH values are marked on the left. On the right are given the excitabilities by induction shocks, in terms of the normal value at pH 7.4. (Lehmann, 1936.)

In mammalian nerve, J. E. Lehmann has recently observed the effects produced by changes in the hydrogen or calcium ion concentration. As in frog nerve, the effects of the two agents are alike. When the pH increases (or the calcium ion concentration decreases), the negative after-potential decreases, and the positive potential increases and shortens (Fig. 85). At the same time the nerve becomes more irritable. At pH 8.1 it is three times as irritable as it is normally at pH 7.4. Between pH 7.6 and pH 7.7 spontaneous discharges start up, and at pH 8.1 a well-marked rhythm appears—all these in a completely reversible manner. Above pH 7.6 the nerve is obviously in tetany.

Demarcation Potential and the Negative After-potential

When the agents bringing about changes in the negative after-potential are classified it is found that substances are brought together which would not be associated for any other reason; and therefore interest is aroused as to what may be the basis of the association. These agents all have some effect on the resting or demarcation potential, and consequently the directions of the change produced in the demarcation potential have been tabulated side by side with the changes produced in the negative after-potential, in order to find out whether or not the variation of the after-potential can be correlated with the state of the fiber surface. The attempt to find a correlation has resulted only in finding a lack of correlation.

A decreased demarcation potential goes with increased negative after-potential in the case of veratrine, and with decreased after-potential in the case of potassium or asphyxia; while increased demarcation goes with increased after-potential in the case of calcium, and a decreased after-potential in the case of narcosis. This lack of parallelism must have a significance of its own. The effects observed are not to be explained by the resting potential—or by the state of polarization or of permeability, as the same idea is often expressed without conveying additional

TABLE IV

	Demarcation potential	*Negative after-potential*
Increased calcium	+[1] −[2]	++[3]
Increased potassium	− −[1]	− − −[3]
Narcosis	+[2]	−[4]
Veratrine	−[2]	+++[5]
Anelectrotonus	+	+[6]
Cooling	−[7]	−[8]
Asphyxia	−[9]	−[10]

[1] Höber and Strohe.
[2] Bishop (1932).
[3] Graham (1933).
[4] Graham (1930).
[5] Graham and Gasser (1931).
[6] Graham (personal communication).
[7] Verzár.
[8] Gasser and Erlanger (1930).
[9] Gerard (1930).
[10] Amberson, Parpart, and Sanders.

meaning. It must not be forgotten that the potential is but the sign of an alteration, and that potentials of the same sign and value may be produced in a number of different ways. It is the nature of the alteration, not its potential sign, which determines the behavior of the after-potential. A simple formulation in terms of the membrane hypothesis is unjustifiable. All that can be said about the after-potentials in terms of this hypothesis is that they represent a process of an unknown nature which affects the condition of the plasma membrane. The same statement can be made concerning the spike. And there the subject must rest until the ingenuity of chemists rises to heights sufficient to permit them to remove the obscurity which surrounds the sequence of metabolic events occurring in the course of an action.

Theoretical Composition of the Action Potential

In order to give a better picture of the sequence of nerve potentials, a diagram (Fig. 86) has been constructed to illustrate how a number of constituent potentials might add together to give the algebraic summation of potentials seen in nerve records. The diagram is a derivation prompted by the results of the various analytical experiments, and its tentative nature must be emphasized.

The schema depicts no particular form of fibers. With variations in the dimensions of the several parts it could be made to fit any of the various kinds of fibers in mammalian or frog nerves and their alterations under changing experimental conditions. Because of the independence of

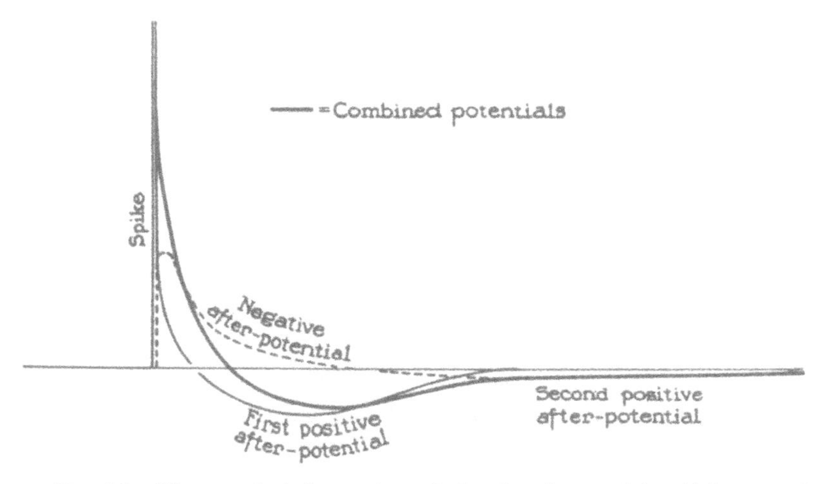

FIG. 86. Diagram depicting a theoretical series of potentials which summed together would reproduce the action potential.

the spike and the negative after-potential, two negative parts are included, and the positive potential is divided to take care of the observation that the positive potential following a tetanus is in two parts. In keeping with the transition from spike to positive potential in C fibers, without an intervening negative potential, and with the increase in the early positive notch after a train of spikes in A fibers, the spike in the diagram is drawn so that it could readily be made continuous with the first positive component. And in keeping with the prolonged positive potential following the augmented negative potential in veratrinized nerve, the negative after-potential and the second positive component are also drawn so that they could easily be made continuous. Continuity would mean that a sequence of events occurs which in its early stages produces a negative potential and in its later stages a positive one.

V

THE EXCITABILITY CYCLE

IN the course of their normal functioning, both sensory and motor nerve fibers are subjected to a wide range of intensities of activity. Motor nerve fibers may be taken as an example. At one time they are called upon to carry only the indolent stream of innervation necessary for the maintenance of postural contraction of the muscles, while at other times, during the performance of rapid skilled movements such as piano-playing, they become the seats of torrents of impulses, now starting, now stopping, as the innervation is shifted from one muscle to another. With all this gradation of activity there must go fluctuations in excitability; for, as we shall see, the momentary excitability of a nerve fiber varies greatly with the intensity of the previous activity. Depending upon the conditions, the excitability may be greater or less than it would be were the nerve fiber permitted to have a long period of rest.

During recent years our notions of the changes in excitability brought about by activity have undergone considerable modification. Frog A fibers, which in the past have supplied most of our knowledge concerning the subject, are now much better understood than they formerly were; and studies on mammalian A fibers and on C fibers of both forms have brought to light properties of the recovery cycle sufficiently different from those obtaining in frog A fibers to necessitate a recasting of the general picture of the process. In the presentation of this picture the information derived from frog nerve must be drawn upon extensively; but, wherever possible, mammalian nerve, which has the greater interest, will be made the basis of the discussions.

The simplest experimental condition that can be set up is one in which only two nerve responses are involved. An initial or conditioning shock is applied to the nerve, and the alteration in the excitability evoked during the course of the response is tested by a second shock. The excitability is then described in terms of the reciprocal of the strength of the testing shock at threshold. Recovery may also be measured in terms of the size of the response, the testing shock being sufficiently strong to bring all the fibers into activity.

The Refractory Period

When activity has started in an A fiber, no additional activity can be brought out by a shock of any strength until the spike has been nearly completed (Adrian, 1921), that is, until the period of rapid restoration following the maximal negativity has come to an end and there remains only the portion of the negativity which is found in the tail. In mammalian A fibers, as seen in records of single axons at body temperature, the major portion of the spike is completed in 0.4 to 0.43 msec. (Fig. 87); hence the absolutely refractory period should last for a like interval.

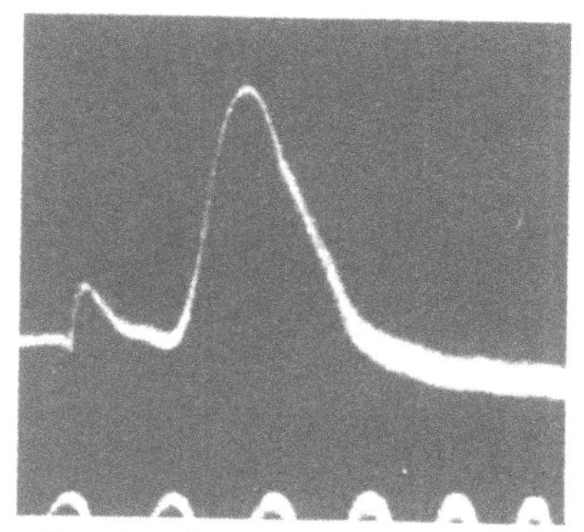

FIG. 87. Spike potential in a single fiber in a sensory root of the cat. The small wave which precedes the spike is the shock artifact. Time 0.2 msec.

Refractory periods lasting between 0.41 and 0.44 msec. are indeed obtained from the best preparations; but not infrequently a second response can be evoked only when the shock interval is prolonged to 0.5 msec. At the first return of responsiveness, the spike is small and its conduction in the but slightly recovered nerve is very slow; therefore, at any point at a distance from the stimulating electrodes the interval between the responses is considerably longer than the interval between the shocks. As the interval between the shocks is increased, both the size of the response and its rate of conduction increase rapidly. This fact brings it about that in the early part of the recovery curve are to be found intervals of responses with spikes of different heights corresponding to them, the differences depending upon whether the interval is made up of a small separation of the shocks and slow conduction, or of a greater separation of the shocks and faster conduction.

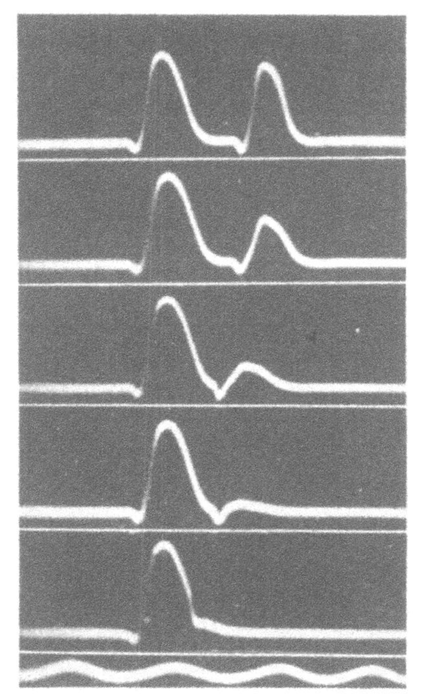

FIG. 88. Relatively refractory period of the phrenic nerve mapped in terms of the height of the response evoked by a supramaximal testing shock. The separation of the conditioning and testing shocks can be determined from the shock artifacts. At the interval of the lowermost record the nerve is completely refractory, but a small response is present in the next record. Conduction distance 8 mm. Temp. 38° C. Time, msec.

The interval between the earliest response and full recovery of the nerve is known as the relatively refractory period. During this interval the threshold for excitation of the most irritable fibers—which are the fibers tested—falls from the high value obtaining at the end of the absolutely refractory period to normal; and at the same time the magnitude of the response increases from its initial small value to full size.

The threshold for mammalian A fibers, in their normal state within the body, typically returns to normal in approximately 3 msec. after the start of the spike; and restoration of the height of the spike should occur at the same time. As the necessary measurements have not been carried out on nerves *in situ*, a definite statement regarding the course of the recovery of height cannot be made; and even the observations made on isolated nerves are such as to give values for the sum of all the fibers, rather than for the fibers in which the recovery of threshold is known. However, the rates of recovery found are so close to what might be expected from the behavior of the thresholds, that the group cannot fall far short of being representative of threshold fibers. In the phrenic nerve, 85 to 90 per cent recovery is present at the end of 1 msec. (Fig. 88).

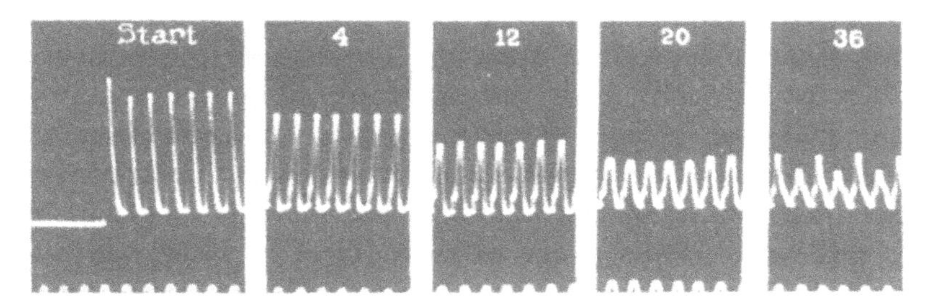

FIG. 89. A train of spikes at a frequency of 1,075 per second. In turn are shown the start of the train and samples of the conditions obtaining after the number of seconds indicated at the tops of the records. The second spike has 91 per cent of the height of the first. During a short introductory phase the spike heights increase slightly, then a long decline in height commences. After 20 seconds alternation begins to be evident, and at 36 seconds it is marked. Phrenic nerve of the cat in 5 per cent CO_2 and 95 per cent O_2. Temp. 37.5° C. Time, msec.

This rate of recovery means that nerves should be capable of conducting 1,000 nearly full-sized responses per second; but tests made on isolated nerves at least show that so high a standard of performance cannot be maintained except for a short period. A progressive falling off in height of the responses occurs, and eventually partial alternation sets in (Fig. 89). Frequencies faster than 1,000 per second involve a still more serious encroachment upon the relatively refractory period, with an attending more rapid decrease of successive responses. Under the condition, however, that conduction is not necessary and that a stimulus stronger than the physiological one can be used—a condition which can be fulfilled at the stimulating cathode—fibers will respond at rates as high as 2,000 per second.

The frequency of innervation of the motor units of muscle is ordinarily only 5 to 50 per second, depending upon the intensity of the muscular contraction (Adrian and Bronk, 1929), and in extreme conditions it does not rise above 100 per second. In sensory fibers the frequencies fall within the same range. For example, Adrian (1932, page 35) found that pressure upon the pad of a cat's foot led to frequencies in a single fiber varying from 9 to 100 per second, depending upon the firmness of the pressure. Only occasionally is it necessary for fibers excited in a physiological fashion to carry impulses at frequencies beyond this range. (Fibers from asphyxiated stretch-afferent receptors carry 450 impulses per second, according to Matthews.) Thus a comparison of the ordinary physiological demands made upon nerve and its capacity to transmit impulses shows that functioning takes place in accordance with the general physiological rule, with the allowance of a large factor of safety.

The Supernormal Period

The end of the relatively refractory period is usually placed at the point at which excitability, spike height, and conduction time return to normal. However, we shall now see that the condition reached at that time is not maintained, but that further changes take place in these properties before the final steady state is reached. Adrian and Lucas established the fact that the relatively refractory period of isolated frog nerve is followed by a supernormal period; and later, association of supernormality with the negative after-potential was shown by Gasser and Erlanger

(1930). Recently H. T. Graham (1934) has examined the nature of this association in some detail.

In a freshly isolated frog nerve with a minimum of after-potential the relatively refractory period ends after 7 to 10 msec., and there is no supernormality (Fig. 90). During the course of experimentation, as the negative after-potential grows, the end of the relatively refractory period comes earlier; but excitability increases for the same length of time as before, and thus the maximum of excitability gets to be definitely greater than the excitability at rest. Supernormal excitability then continues during the progress of the negative after-potential, and at the same time conduction becomes supernormal in velocity.

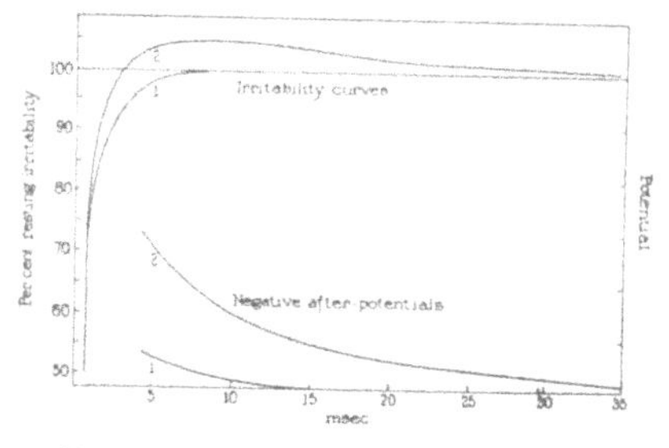

FIG. 90. Recovery curves of a green frog sciatic nerve compared with the size of the negative after-potentials. 1. Soon after the nerve was mounted. 2. After about an hour of testing thresholds. The resting excitability and the spike height were constant throughout. Abscissae: interval between conditioning and testing shocks. Ordinates calculated from the strength of shock necessary to excite at threshold. Testing shock applied to the nerve 17 mm. from the conditioning shock. Temp. 26° C. The increased negative after-potential shortened the relatively refractory period from 9 msec. to 3 msec. (Drawn from H. T. Graham, 1934.)

All the factors controlling the rate of recovery are not known, but the evidence at hand indicates that at least two processes are involved. Part of the control seems to belong to a process with a relatively fixed time course. The absolutely refractory period, and to a less degree the time required to reach maximal excitability, are not easily changed. Shortening of the relatively refractory period to one quarter and the development of a long supernormal period may occur without alteration of the absolutely refractory period or change in the time during which maximal excitability is attained. As the spike is known to have a very constant shape, it may be inferred that the spike process plays the dominant part in determining the duration of absolute refractoriness and the time of rising excitability. Except in instances in which the crest of the negative after-potential is so long delayed that it comes beyond the range of influence of the spike, the maximum of excitability does not correspond to the maximum of potential.

On the other hand, the very variable after-potential process may be considered as being responsible for the increased rate of recovery which shortens the relatively refractory period, and also for the extension of the period of supernormality.

The Subnormal Period

In frog A fibers the supernormal phase is the last measurable event before the return of the fibers to normal excitability at the end of a single response; but this mode of ending is not typical of nerves in general. The first evidence for a different ending was found by H. T. Graham in the recovery curve of yohimbine-poisoned nerves. Nerves poisoned by this drug or by a number of others which imitate it in a less striking manner, displayed a second period of subnormal excitability following the supernormal period, now curtailed in duration. This subnormal excitability was soon correlated with the positive after-potential (Graham and Gasser, 1934) which proved to be increased in magnitude; and it was also found that subnormality could readily be demonstrated in unpoisoned nerve in the period of augmented positivity following a tetanus.

The subnormal excitability which occurs during the positive potential is accompanied by a subnormal velocity of conduction, but the spike remains full-sized. The subnormal period differs in this regard from the relatively refractory period in which the spike is undersized. Were it not for this fact, one might suppose that the relatively refractory period and the subnormal period are one continuous event in which is interpolated a supernormal period associated with the negative after-potential.

The existence of a subnormal velocity of conduction during the subnormal period is strikingly brought out through the effect of the slowed conduction on temporal dispersion. Slowing of conduction in all the fibers of a nerve, even when it is not differential, increases the separation of the secondary crests of the action potential. Thus alternation between normal and subnormal conduction should be marked by alternate approximation and dispersion of the crests. A reversible change of this sort occurs between conduction in rested nerve and conduction in the period following a tetanus (Fig. 91). In the form of the action potential obtained during the latter there is a delay in the time of arrival at the recording electrodes of the impulses in the largest fibers, and there are progressively greater delays of the impulses in fibers of successively smaller sizes. Accordingly, the beta crest is delayed more than the alpha crest, and the gamma crest more than the beta; and at the same time, because of the greater dispersion of the potentials within the group, each elevation in turn is decreased in height and prolonged to a greater extent than the preceding one.

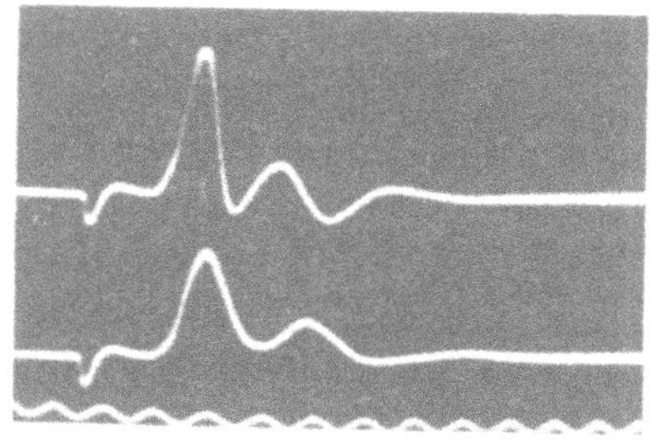

FIG. 91. Temporal dispersion in the A fibers of the sciatic nerve of the bullfrog as it is affected by the subnormal velocity obtaining in the subnormal period. The upper curve is taken from a rested nerve, the lower one during the positive after-potential following a tetanus. The testing electrodes were between the leads and the tetanizing electrodes, so that the local effects produced by the latter would not affect the conduction time. Conduction distance 6 cm. Temp. 22° C. Time, msec.

None of the phenomena connected with the subnormal period can properly be described as caused by fatigue or

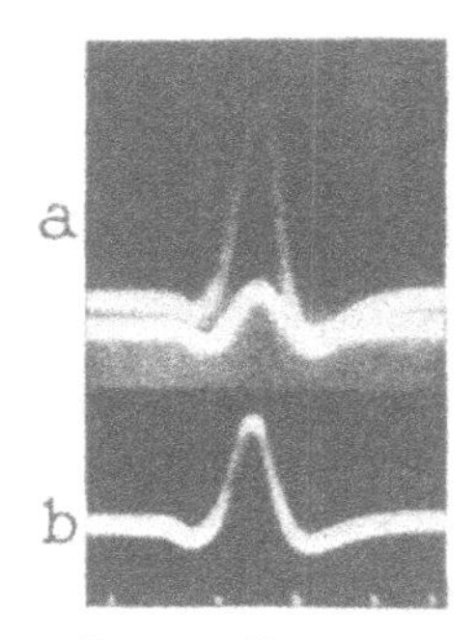

FIG. 92. Temporary change of the excitability during the subnormal period from subnormal to supernormal by interpolation of a single response. The testing spike was evoked by a weak shock which stimulated only a small group of near-threshold fibers. Its size in rested nerve is shown by *b*. Part *a* is made up of two superimposed records. The lower, smaller one is the response to the testing shock during the subnormal period following a tetanus. The upper record is in the next succeeding sweep. Just before the sweep a maximal response was interpolated. Owing to the negative after-potentials of the latter, the sweep is at a higher level (D.C. amplification) and the size of the response now evoked by the testing shock shows that excitability is supernormal. Time, msec.

exhaustion of the nerve. If during the subnormal period following a tetanus a shock sufficiently strong to stimulate is interpolated, the excitability is raised to supernormal during the negative after-potential of the interpolated response (Fig. 92). While this supernormality is only temporary and is in truth nothing more than the supernormal period of a single response developing from a subnormal instead of a normal level, the fact that the development can take place shows the presence of capacity for supernormal responsiveness, ready to be called upon should demand be made for further intense activity.

Mammalian A fibers have a recovery cycle possessing qualitatively the same features as frog A fibers; but a well-marked subnormal period after single spikes, corresponding to the clean-cut positive after-potential, sets them in contrast to the latter. The definite subnormal period, in fact, links them to C fibers of both forms, in which subnormality is particularly marked.

Excitability of C Fibers

The earliest second responses in C fibers, as in A fibers, start at the base of the spike and also, as in the latter, are initially small and increase rapidly in size as the interval between the responses is made longer (Fig. 93). The second spike tends to reach at its crest the absolute negativity holding for the crest of the first spike; and, therefore, inasmuch as it rises from a level of positive potential, its height over all soon comes to be greater than normal. C fibers differ in this regard from A fibers.

Throughout the period in which the spikes are supernormal in size the excitability is subnormal. When the whole course of the recovery is plotted out by the method of thresholds, as has been done for frog nerve, it is found to possess a very interesting feature. Beginning at the end of the absolutely refractory period, the excitability rises rapidly to normal; then, instead of holding its value or becoming supernormal, it declines again and remains subnormal until the positive after-potential has been dissipated (Fig. 94). Refractory period and subnormal period merge into one long unresponsive period, the course of which does not follow a smooth curve by reason of a small temporary increment of excitability. This increment may be interpreted as corresponding to a negative after-potential process too small to give either a detectable negative after-potential or a period of supernormality. If the process be developed with veratrine, then after-potential and supernormality appear in abundance. All during the time in which the excitability had been subnormal, it now becomes supernormal (Fig. 95); and subnormality is dis-

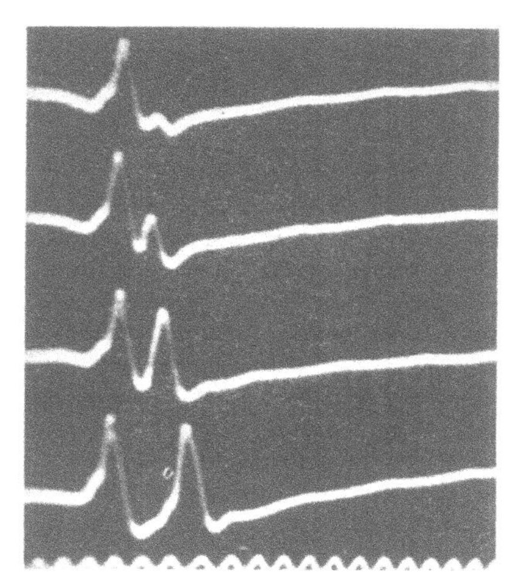

FIG. 93. Recovery of height of response in C fibers. Splenic nerve of the cat at 37° C. Supramaximal induction shocks are applied at 4 intervals. Conduction distance 8 mm.; time, 5 msec. As the earliest responses in the relatively refractory period are greatly delayed in C fibers, the shock interval in the uppermost record is much shorter than the response interval.

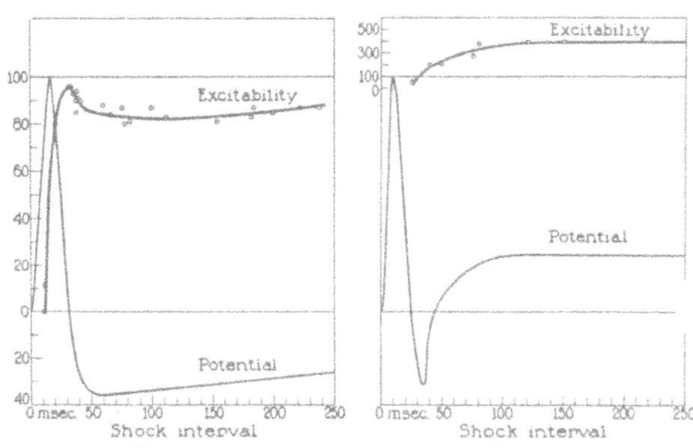

Fig. 94. Fig. 95.

Fig. 94. Recovery curve of C fibers, splanchnic nerve of the bullfrog. The excitability is determined as the reciprocal of the strength of shock necessary to elicit a response and is reported on the ordinates as per cent of the normal threshold. The form of the action potential is also plotted. The excitability curve cuts the spike curve because of the prolongation of the latter by temporal dispersion, which is large even at 4 mm. of conduction. The axon C spike is not longer than the absolutely refractory period. Temp. 23.5° C. (Richards and Gasser.)

Fig. 95. Recovery curve of the same nerve that is recorded in Fig. 94, after the latter had been treated with veratrine. The high degree of supernormality which has developed will be noted.

placed to a later period which awaits the subsidence of the negative after-potential (Fig. 96).

Mammalian C fibers undoubtedly behave like frog C fibers. Although the curves have not been drawn, subnormal and supernormal excitabilities of the same character have been demonstrated (Fig. 97).

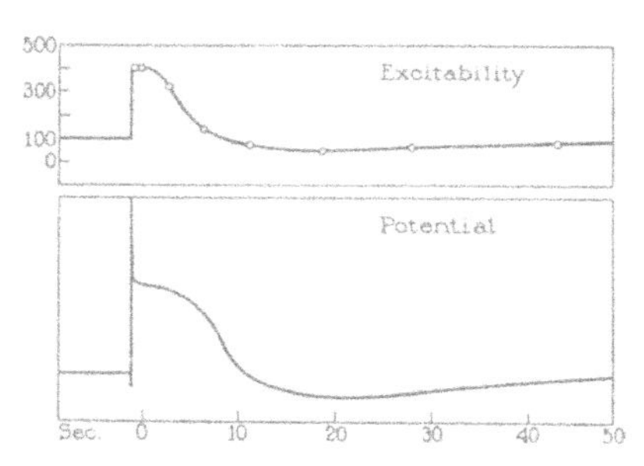

Fig. 96. Continuation of the excitability curve recorded in Fig. 95 showing that, associated with the positive after-potential, a long period of subnormality follows the period of supernormality.

Excitability of Mammalian A Fibers in situ

After becoming acquainted with the course of the recovery curves of C fibers, the curves obtained from mammalian A fibers in their normal physiological state within the body will not seem to be exceptional to what holds for nerve fibers in general. The curves are characterized by a rapid rise to maximum, a decrease from the maximum, and then a long subnormal period.

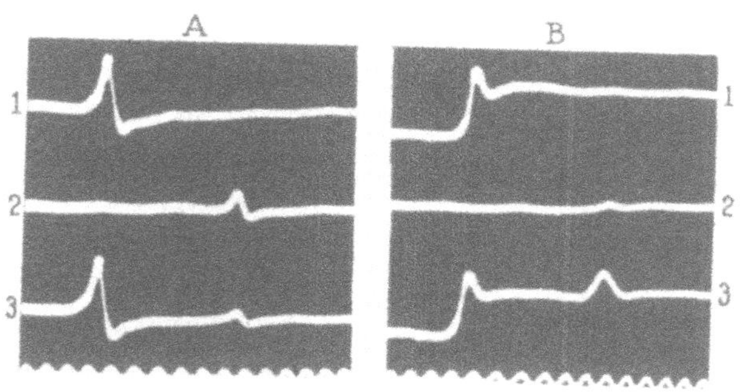

FIG. 97. Supernormality and subnormality in mammalian C fibers. Splanchnic nerve of the cat. Temp., 37° C. Time, 5 msec. A 1. Normal action potential at 8 mm. of conduction. A 2. Control height of a small testing response. A 3. Response to the same shock during the subnormal period. B 1. Action potential after the nerve had been treated with veratrine, 1 : 500,000. B 2. Control height of testing response. B 3. Response to the same shock during the supernormal period.

When the conditioning action is single, the maximum is as a rule at a supernormal level (Fig. 98). Supernormality begins at approximately 3 msec., at the end of the relatively refractory period, and increases to a maximum which comes between 3 and 10 msec. The supernormality then decreases and ends when subnormality begins, at 12 to 18 msec. Maximal subnormality is reached at 25 to 35 msec., and normal excitability is finally restored at 60 to 80 msec.

When the conditioning action is a short tetanus, the maximum no longer rises above 100 per cent excitability, and the curves now resemble those obtained from C fibers conditioned by single responses. As the conditioning excitation is made more intense through a prolongation or a greater frequency of the tetanus, all parts of the curve are lowered and the total period of subnormality is prolonged. Subnormality reaches a greater maximum and the maximum comes earlier (Fig. 98)—changes which are quite in keeping with the deepening of the first positive notch of the after-potential and the more rapid decline of the negative after-potential (Fig. 79). An earlier attainment of maximal subnormality means a curtailment of the duration of the initial peak rise of the excitability; and at the end stage of this progression the peak is obliterated altogether (lower curve, Fig. 98). The period of rapid rise of excitability characteristic of the relatively refractory period then becomes arrested at a level far short of normal, and from that point onward recovery takes place slowly. After a conditioning tetanus of one minute, this recovery period has been found to last one minute.

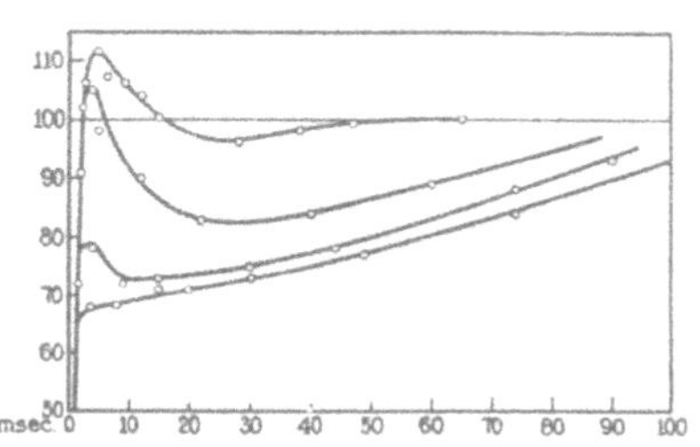

FIG. 98. Recovery curves, cat saphenous nerve *in situ*. Abscissae, separation of conditioning and testing shocks. Ordinates. Excitability measured as the reciprocal of the strength of the stimulating shock at threshold and reported in per cent of the threshold of the resting nerve. The upper curve is conditioned by a single shock, the next one by 3 shocks at 90 per sec., the third by 7 shocks at 250 per sec., and the lower curve by 23 shocks at 250 per sec. (Gasser and Grundfest.)

The excitability curves of isolated nerves differ from those of nerves *in situ* largely because of the greater negative after-potential which is found in isolated nerve. The supernormal period after a single response is greater; and after a tetanus supernormality may appear a second time at the end of the subnormal period. In that case there is a final return to subnormality, as would be expected from the course of the after-potential in such nerves. In short, the excitability follows the after-potential; if the latter is oscillatory, the excitability oscillates with it.

Significance of Supernormality

The question may well be asked, what has supernormality to do with the functioning of nerve fibers? Nervous activity usually involves a tetanus, and as we have seen, after a tetanus supernormality does not occur. This does not mean, however, that the effect of the process is not present. The early part of recovery is undoubtedly accelerated by it. After a mild tetanus, the early maximum of excitability is as unmistakably caused by the process of supernormality as if actual supernormality were present; and even after a severe tetanus, when an early peak is no longer present, the process is in all probability effective in raising the level at which the early rapid rise of excitability becomes arrested.

But, whatever may be the importance of the supernormality process in bringing about recovery after a tetanus, a greater importance may be ascribed to it during the tetanus. Responses at high frequency are facilitated at no greater cost than a postponed period of subnormality, from which the nerve can be called back promptly into readiness for action through the interpolation of a response and its subsequent lowering effect on the threshold (Fig. 92). Without the process of supernormality—or what is the same thing, the negative after-potential process—the power of sustaining tetani is greatly reduced. If, for instance, a nerve be brought by mild asphyxia to a state in which it produces good single spikes, but poor negative after-potentials, a train such as that shown in Figure 89 dies out very rapidly.

Raising the excitability temporarily, either from the resting level or an equilibrated subnormal level, supplies an excellent mechanism for picking up excitation from a subthreshold background. How the mechanism operates can be illustrated by a simple experiment. A mammalian nerve is stimulated by thyratron shocks at a rate somewhat above 100 per second and at a strength just at threshold for the most irritable fibers. To this excitation is now added a single shock sufficiently large to stimulate the whole nerve (Fig. 99). In the supernormal phase of the ensuing response the thyratron shocks become suprathreshold for all the fibers in the nerve, and thus a second maximal response is produced. Then successive thyratron shocks fall in successive supernormal phases, and excitation of the nerve continues until curtailment of the supernormal phases by the growing tendency to produce positive potential (Fig. 79) causes the fibers to fall below threshold within the compass of the interval of stimulation. This they do differentially in the order of their natural thresholds; and, once they have dropped out, they cannot be picked up again, as their excitability continues to decrease and becomes subnormal. Thus the nerve undergoes a long self-limited discharge of impulses.

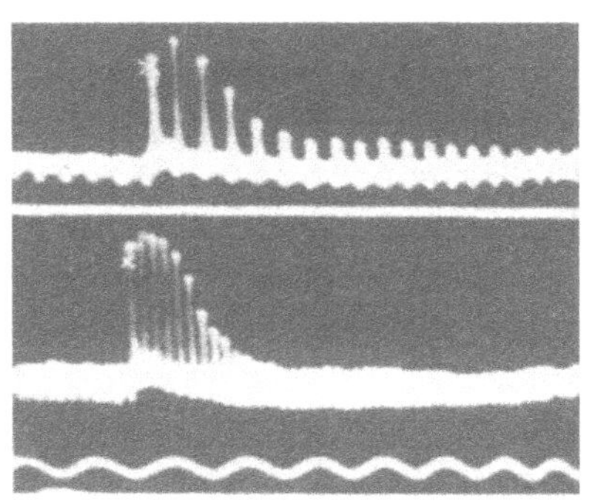

FIG. 99. The effect of a single maximal response in a nerve receiving tetanic subthreshold excitation. Isolated phrenic nerve of the cat, 37° C. Shocks are applied rhythmically from a thyratron stimulator. In the first part of the sweeps can be seen the responses evoked from the equilibrated nerve. Very few fibers are responding, the rest having too high a threshold. The deflections below the line are shock artifacts. At *x* a single maximal induction shock is applied to the nerve at a point more removed from the leads than the thyratron shocks. It will be noted that not only are all the fibers brought into action in the supernormal phase of the added response, but that there is staircasing of the negative after-potential as well (compare Fig. 78). Upper line: frequency 135 per second. Lower line: frequency, 325 per second. Time, 20 msec.

Relation of the Irritability of Neurones in the Spinal Cord to their Potentials

A long-lasting train of responses initiated by an isolated disturbance has so great a superficial resemblance to some of the events occurring in the central nervous system as to render imperative an examination of the latter with a knowledge in mind of the factors effecting control of excitability in peripheral nerve. The specific question is raised: is the irritability of central neurones modified by processes the electrical manifestations of which may properly be called after-potentials? *A priori* there is good reason to expect that after-potentials would be present in the central nervous system. They occur in all types of peripheral axons, and if the latter may be taken to be representative of nervous tissue in general, the presence of after-potentials in the central ganglia would logically follow.

Although studies of the central nervous system from the standpoint of the after-potentials have been limited in number, there are available a few observations which can be cited as bearing on the problem. No potential has as yet been definitely identified with a negative after-potential, nor has supernormality been described; hence, even though the possibility of the presence of these two events is by no means excluded, it is at this time unprofitable to extend the argument further in that direction. On the other hand, there is evidence that processes occur not only in the central nervous system (Gasser and Graham, 1933), but also in the retina (Granit, 1933) and in sympathetic ganglia (Eccles, 1935), which have potential signs resembling positive after-potentials, and which are accompanied by the excitabilities appropriate to this potential.

Before entering upon a discussion of the potentials in the central nervous system, or more specifically in the spinal cord, it will be well to consider schematically the neurone linkages which lie in the path of a simple spinal reflex, such as the homolateral flexor reflex. On purely histological evidence there is reason to believe that even in the simplest reflexes at least three neurones are involved. E. C. Hoff has shown that after the dorsal roots of the cat are sectioned between the ganglia and the cord, the endings which degenerate are in relation to internuncial neurones, rather than to motor neurones—which means that excitation would have to be relayed before the motor neurones are reached. This fact should not, however, be taken to mean that there are no two-neurone arcs. Cords treated with stains coloring the myelin show quite the contrary to be true. Very large fibers have been described by both Cajal and Winkler as passing directly from the dorsal columns to motor cells.

Three locations present themselves in which cord potentials might arise (Fig. 100): (1) the endings of the primary neurones; (2) the internuncial neurones; and (3) the cell bodies and dendrites of the motor neurones. It would be expected that the endings of the primary neurones would have the same potentials as the fibers supplying them, as presumably they are but divisions of the same kind of protoplasm. In the terminal arborization of the neurones, slower conduction would occur, because of the diminution of diameter which is found at each division of the fibers; but the effect of the slowing would not be apparent because of the shortness of the branches.

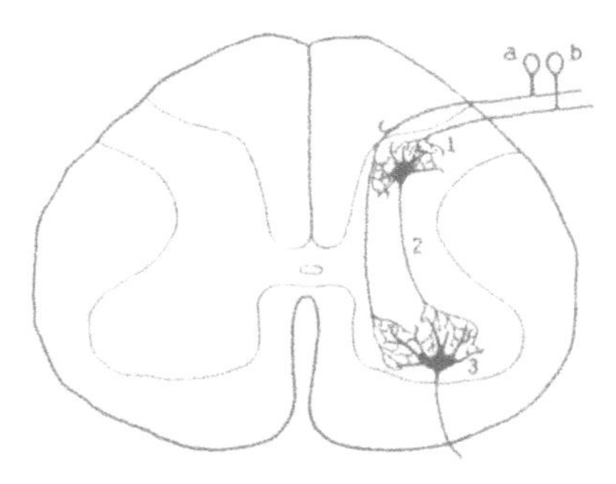

FIG. 100. Diagram of the spinal cord showing in simple form a few of the connections involved in an ipsilateral reflex.

The potentials in motor cells cannot be predicted from peripheral nerve. Histologically a differentiation occurs at the axon hillock between the appearance of the cytoplasm of the perikaryon and that of the axon. If the cytoplasm were everywhere alike, the only variation which the cell body might be expected to introduce would be one associated with its dimensions. With histological differentiation, on the other hand, the possibility exists that chemical processes differing in nature from those in the axon may take place and that consequently the potential sign of the processes may be different. If, therefore, there are potentials in the cord not completely in accord with those in motor axons, the motor cells must be considered.

Also, nothing is predictable about the internuncial neurones. They might have a potential differing from peripheral A fibers because of the presence of the cell body, or because the whole neurone is different. Just as greatly differing axons are to be found in peripheral

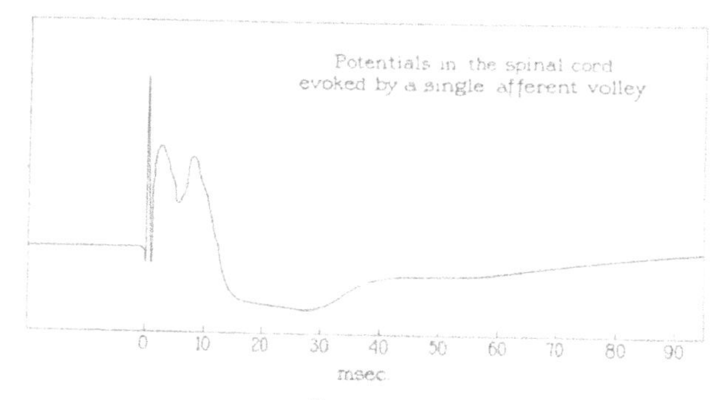

Fig. 101.

nerve, so it may be anticipated that a variety of neurones is present in the nervous system. In fact, we know that there is more than one variety, because some of the neurones send out dissimilar axon processes into the nerves.

After this preliminary discussion, we may consider what actually happens when an afferent volley comes into the spinal cord. If the potentials be led off through two electrodes on the surface of the cord, as they occur when a single induction shock is applied to a dorsal root, they show basically first a spike (Fig. 101), then a longer and somewhat lower negative potential (the electrode nearer to the root stimulated is negative to the one farther away), and finally a still longer and lower positive potential. Complications of this simple picture occur, depending upon the condition of the cord and the extent to which excitation spreads or reverberates before it subsides; but they obscure, rather than elucidate, the essential features, all of which are present in the basic form described.

The intramedullary spike is undoubtedly in the central continuation of the dorsal root fibers. It is conducted centrally for a few centimeters at the velocity at which it enters, then conduction is quite abruptly slowed, and the impulse goes on toward the brain at less than half the initial velocity. The most probable explanation of this phenomenon is that the fiber which passes up through the fasciculus gracilis, after all the collaterals are given off to the segmental nuclei, is much smaller in diameter.

The negative and positive potentials which follow the spike are much larger than the after-potentials of sensory A fibers. This fact in itself indicates that the potentials are not in the endings of primary neurones. There is, however, a better reason for this conclusion. If two roots are stimulated, or two branches of the same root, the potentials produced by the components will add as long as they arise in parallel pathways, but they will fail to do so when the pathways converge on to common lines. The intramedullary spikes do as a matter of fact add; but the potentials which follow them do not add. The potentials evoked through simultaneous stimulation of two branches of a root are but slightly larger than those evoked through stimulation of one of the branches alone. This means that the cord potentials arise in units other than those which produce the spikes and that, therefore, they occur beyond the first neurone, with either motor cells or internuncial neurones to choose from.

If the motor cells were the source of the potential, it would be expected that the potential could also be brought out by stimulating the cells antidromically, according to the method of Denny-Brown, that is, by impulses coming to them from the motor axons; for the evidence is good, from the experiments of Eccles and Hoff, that antidromic impulses actually do pass backward over the cells. (In their experiments, rhythmically discharging neurones have their rhythms altered by a volley back-fired into the cord; and an alteration in rhythm such as this would hardly be possible if the impulse did not get back to the point at which excitation of the neurone takes place.) A typical cord potential is not brought out, however, by antidromic excitation. A spike is visible and there is also evidence from excitability data that a positive after-potential may be present; but the latter is not seen, probably because a potential of the dimensions of the one on motor fibers is tco small to be detected under the experimental conditions. There is no positivity which resembles the positive potential elicited by volleys arriving in the normal direction. Activity in motor neurones, therefore, cannot be the cause of the large negative and positive potentials seen when the cord is excited by an afferent volley.

Subthreshold excitation has been suggested as the cause of the potentials, but with all the current evidence indicating that there is no potential accompaniment of the phenomena of subthreshold excitability sufficiently large to be detected, this view cannot be used in argument, if we are to confine ourselves rigidly to the citation of phenomena known to occur in peripheral nerve and of properties of neurones experimentally demonstrated. Thus, by exclusion, the internuncial neurones become the most likely locus of the cord potentials.

The general form of the cord potential at once strikes one as being similar to that of peripheral nerve; and from this resemblance the most satisfactory hypothesis for the interpretation of the potential may be built. The unit internuncial neurone potentials may be thought of as consisting of a spike (possibly associated with some negative after-potential) followed by a positive after-potential. The spikes of these units temporally dispersed would then give the negative part of the composite cord potential (in Fig. 101 in two major groups), and the ensuing positive after-potentials the positive part. On account of the short duration of the spike and the long duration of the positive after-potential, temporal dispersion would curtail the summed height of the spike portion much more than that of the after-potential portion, and thus the large size of the latter in comparison with the former, which is in contrast to the relationship obtaining in records of peripheral A fibers, would in part be accounted for. It is also possible that the positive potential of the cord neurones may be larger with respect to the spike than it is in A fibers, in this regard resembling C fibers more closely than A fibers. All the more complex potential forms encountered in highly irritable preparations would also be accounted for, if spread of excitation to additional neurones and the possibility of the occupation of reverberating chains in the manner described by Ranson and Hinsey are brought into the argument.

The hypothesis may be tested by the application of two afferent volleys in succession. If the negative cord potential is made up of spikes, the neurones generating it should be unresponsive to further excitation as long as the potential lasts.

Before a description of the test is presented it must be recalled that the strength of the stimulation received by an internuncial neurone depends upon the number of active endings on its surface. Some neurones receive but a few endings, others many. Accordingly, as has been pointed out by Sherrington (1931), some neurones must be stimulated just at threshold, and others at various strengths above threshold. Complete responses will not be evoked by the second of two volleys until all the neurones have recovered to the point of responding to the strength of stimulation which the endings can deliver; therefore, the recovery curve will belong to the type obtained with weak stimuli, and for that reason should be compared with recovery curves of peripheral nerve mapped with weak shocks.

We have seen that when the testing shock is weak, C fibers and A fibers conditioned by a tetanus have recovery curves marked by a rapid initial course, a prolonged final course, and no intervention of supernormality (Figs. 94 and 98). We shall now see that when allowance is made for the properties of the cord which militate against getting the true recovery curve of single neurones, the recovery curve of the negative cord potential has the same general trend.

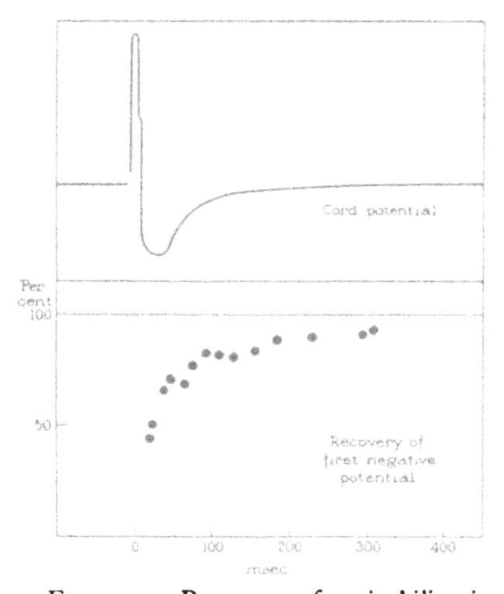

FIG. 102. Recovery of excitability in internuncial neurones. The upper curve shows the form of the response evoked by the isolated conditioning and testing volleys. The lower curve indicates the per cent of the normal response which could be brought out by the testing volley when it followed the conditioning volley at the intervals marked in the abscissae. By the first negative potential is meant the first peak of the cord potential. The part of the negative potential which follows the notch on the down stroke recovers later. (Hughes and Gasser.)

Excitability of the spinal cord is measured by a method similar to that used for measurement of the excitability of nerve. A stimulating shock to a dorsal root or sensory nerve produces a volley of impulses which conditions the cord, and a second volley measures the responsiveness after conditioning. No significant additional potential can be elicited by a testing volley until the volley falls after the negative portion of the response to the conditioning volley. Then it brings out a small amount of negative potential (Fig. 102). As the interval between the conditioning and the testing volley is lengthened, the area of the added negative potential increases. The increase is rapid at first; but the rate of increase falls off in the later portion of the curve, and full-sized responses are not obtained until the positive phase of the conditioning response has subsided. The negative potential thus behaves as though it were produced by spikes. If negative after-potential were an important part of the potential, spikes corresponding to antecedent spikes should be readily re-elicited, but this does not happen.

Discharge of Motor Neurones

In suitable preparations, throughout the course of the negative part of the cord potential, a discharge into the sciatic nerve occurs (Fig. 103). From an activity in the primary neurones lasting for a millisecond or two there results a motor discharge of 25 msec. According to the observations of Lorente de Nó (1935*d*) made on the nucleus of the third cranial nerve, the local excitatory effect in a motor neurone persists for a period not longer than 0.5 msec. after arrival of nervous impulses at the synapse; therefore, a discharge of 25 msec. must be kept up from an external source. That this source is in the neurones, which produce the negative cord potential, is obvious. The latter may be considered to be a reservoir of

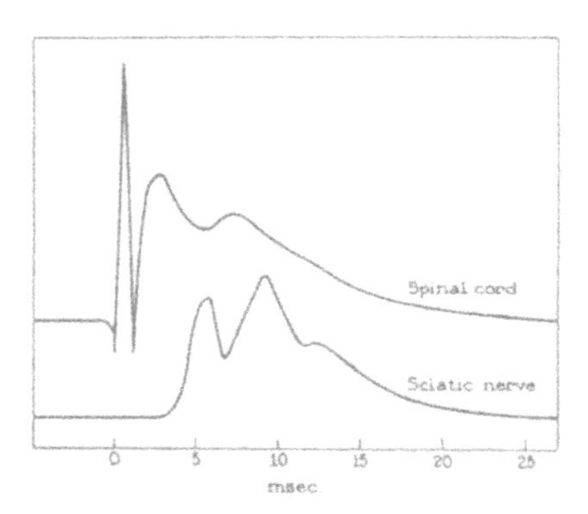

FIG. 103. Comparison of the duration of the discharge of motor impulses into the sciatic nerve with the duration of the negative part of the potential evoked in the internuncial neurones of the spinal cord by a single afferent volley from a dorsal root. The experiment was performed under light ether anesthesia. (Drawn from Hughes and Gasser.)

excitation from which the motor neurones are supplied with a succession of stimuli.

Negative Cord Potential and Facilitation

While the negative cord potential is in progress, facilitation is possible. A reflex evoked by an afferent volley is larger when the volley is conditioned than when the volley is in isolation. Two successive volleys, when both are small, may produce a reflex larger than the sum of the responses to the isolated components.

Facilitation, as it is at present known, is a phenomenon in motor neurones. It has not yet been definitely identified with internuncial neurones.[1] In fact, we have just seen that the internuncial neurones are very unresponsive to a second volley during this period. The explanation of facilitation, therefore, must be stated in terms of the events occurring in and about the motor cells.

The behavior of the potentials in the internuncial neurones suggests what the mechanism may be. In some preparations, facilitation of a flexor reflex is easily produced; in others, the testing response is depressed rather than augmented (Eccles and Sherrington, 1931). The cord potentials corresponding to the two variations are then found (Hughes and Gasser) to be as different in appearance as the nature of the reflexes. It is possible, through observation of the potentials alone, to predict the character of the reflex which will be produced by the testing volley.

When the end result will be facilitation, little positivity appears in the course of the cord response, and considerable

[1] Dr. Joseph Hughes has been so kind as to show me some records made by Dr. Grayson McCouch and himself which apparently prove that facilitation also occurs in internuncial neurones. The necessary condition for its production seems to be making the conditioning volley very weak, and causing it to be followed by a testing volley so early that it falls during the period of latent addition, as described by Lorente de Nó (1935*d*). The finding of facilitation of internuncial neurones serves only to extend the application of the argument which follows and does not upset its fundamental validity.

additional negative potential is evoked by the conditioned volley (Fig. 104 A). This situation suggests that facilitation must be brought about through accumulation of excitation within the motor neurones, resulting from an addition of a fresh influx of impulses to the bombardment residual from the first volley. If the effect of the first impulses to arrive from the conditioning volley be wiped out by an intercurrent discharge of the neurones called forth by antidromic stimulation, as in the experiments of Eccles (1931) and of Lorente de Nó (1935*b*), after the resulting refractory period there is still time for facilitated excitation to be recreated through convergence of impulses responsible to the conditioning volley arriving at the neurones over delay paths with impulses responsible to the testing volley.

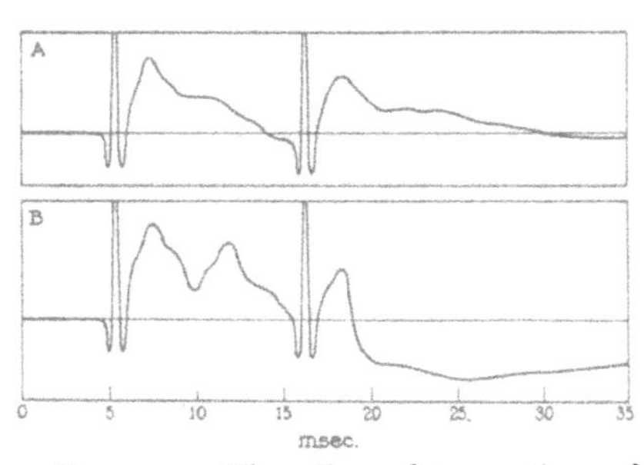

FIG. 104. The effect of two volleys of the same size upon the potentials in the internuncial neurones of the spinal cord. Both curves are obtained from the same preparation, but there is a difference in the experimental state. A. Intact central nervous system, light ether anesthesia. B. After spinal transection at the twelfth dorsal segment, decerebration, and the elimination of the ether. The "B" response approximates more closely to normal than does the "A" response. As both shocks were of the same size, the testing volley in isolation should produce the same response as the first volley. (Drawn from Hughes and Gasser.)

Positive Cord Potential and Inhibition

When the end result will be a depression of the conditioned reflex, there is a large positive potential in the course of the cord response and a marked deficit in the negativity produced by the conditioned volley, as compared with the negativity in a control unconditioned response (Fig. 104 B). Depression as here described is usually called "inhibition" in the terminology of the physiology of the central nervous system. Inhibition is best brought out by causing a large testing volley to succeed a small conditioning volley. The conditioning volley by itself produces but a small reflex, and the two volleys together, even at short intervals of separation, produce a reflex effect much smaller than the testing volley would produce in isolation. A figure from an experiment of this sort is reproduced from Eccles and Sherrington (1931) and may be taken as being typical of many which we have ourselves encountered. The temporal parallelism between this curve and the course of the positive potential in the spinal cord at once strikes the eye.

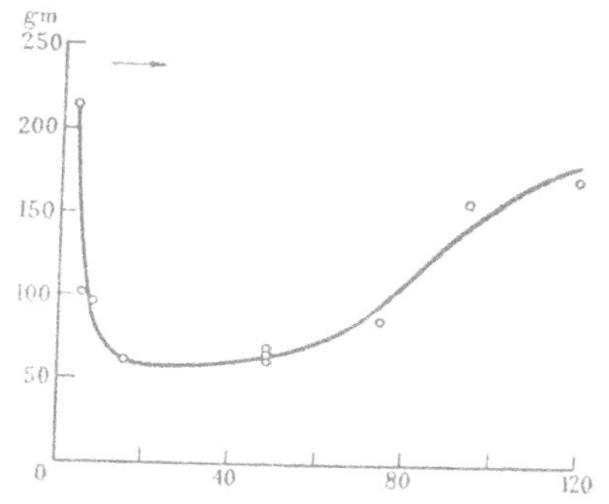

Fig. 105. Maximum tension (ordinates) reflexly evoked in a tibialis muscle by two afferent volleys, plotted against the separation of the volleys in milliseconds (abscissae). The tension of the reflex evoked by the testing volley alone is indicated by the arrow. The conditioning volley was very weak and by itself produced very little tension. (Eccles and Sherrington, 1931.)

That the activity connected with the negative cord potential is indeed essential for the excitation of motor neurones is well brought out in these experiments in which inhibition is prominent. According to the conditions of the experiment, the size of the second response depends largely upon the exciting power of the second volley. As soon as the second response gets sufficiently far away from the first to escape all facilitation by it, any increment of tension produced in a reflexly excited muscle over that produced by the conditioning volley, must be produced entirely through the unaided action of the second volley. The increment may then fall to zero (Hughes and Gasser). At the same time, the negative potential is to a large extent eliminated. An idea of the completeness to which this is possible may be gained by reference to a record obtained in my laboratory by D. Clark (Fig. 106). The potentials were recorded from the unexposed cord through concentric needle electrodes, and the afferent volleys were supplied by the A fibers of the saphenous nerve. But a trace of the proper normal response was brought out by the testing volley.

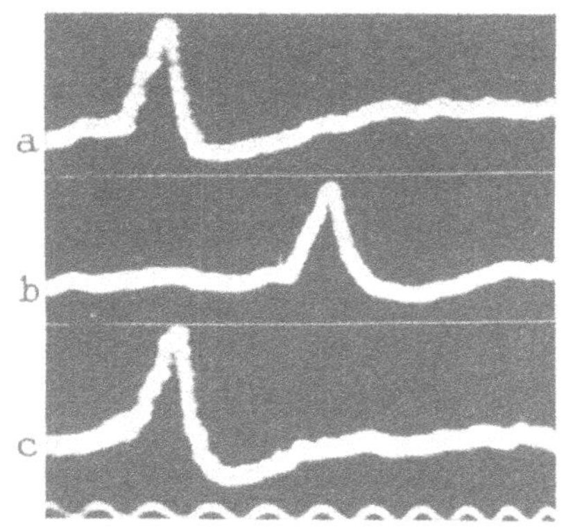

Fig. 106. Depression of a response of the internuncial neurones of the spinal cord by an antecedent response. a. Action potential evoked by the conditioning volley. b. Action potential evoked by the testing volley in isolation. The two responses are the same; but the testing response is placed 60 msec. later in the sweep. c. Action potential evoked by the two volleys in succession. The response to the second volley is nearly obliterated. Spinal preparation. The width of the line is caused by the large amount of background activity. Time, 20 msec.

When the interval of stimulation becomes sufficiently great, reflex muscular contraction begins to be elicited again by the second volley, and its size increases progressively as the latter becomes separated from the conditioning volley. As we have previously seen, the negative cord potential also increases. The two expressions of activity are restored to their normal magnitude together, final restoration in either case not being reached until the positive cord potential of the conditioning response has ended.

Production of a gross deficit in the response to a large afferent volley, by causing a small afferent volley to precede it at an appropriate interval, is a phenomenon which cannot be duplicated in peripheral nerve by causing a weak stimulus to precede a strong one. Failure of duplication, however, does not necessarily mean that any processes present in the central nervous system are missing in nerve.

The difference is to be attributed rather to the arrangement of the units. All nerve fibers are completely independent of one another, while the central neurones are linked together in chains. Activity in a fraction of a nerve's fibers has no influence on the ease with which the other fibers may be excited, but activity in a fraction of the

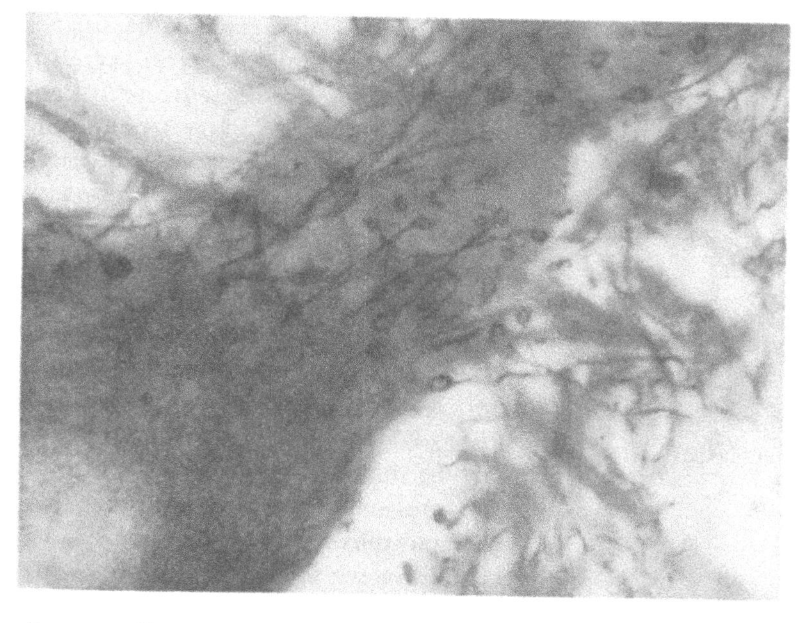

FIG. 107. Endings on a portion of the cell body of an internuncial neurone in the spinal cord. The microtome knife has cut off a part of the cell surface, so that the nucleus is visible in the lower left hand corner. Magnification × 1100. (E. C. Hoff, 1932.)

cord's neurones makes others less easily accessible, as the impulses to the latter must travel over pathways which have acquired a high threshold.

The cells and dendrites of neurones are literally covered with the terminations of axons. Some idea of their density may be gained from a section made by E. C. Hoff, of which he has kindly furnished me a photograph (Fig. 107). The endings come from numerous sources, descending tracts and segmental paths connected with afferent fibers, and excitation may be considered as taking place when a sufficient number of these endings or terminal knobs (*boutons terminaux*) becomes active at one time.

Let us suppose that a number, N, of these knobs must be active for excitation to take place. And for the sake of simplicity in the construction of a diagram let us suppose that the value of N is 2. A simple schematic, synaptic system may then be set up, making use of neurone linkages of types known to exist in the cord, to show how transmission through the cord may be facilitated or inhibited (Fig. 108). In the figure, primary, internuncial, and motor neurones are labeled in turn 1, 2, and 3. Fiber X represents the fibers in a sensory nerve which are stimulated by a weak shock, and fiber Y represents the additional fibers which would be stimulated by a strong shock. When X is stimulated, the line is clear to motor neurone A, and it discharges. The branch from X_1 to B is insufficient to excite, and as the local excitatory process it produces would not last longer than 0.5 msec., according to Lorente de Nó (1935*d*), no summation with the endings from X_2 would occur, because the impulses in the latter are delayed by a synapse time of 1 msec. at X_1 to X_2. Two volleys over X, however, if spaced 1 msec. apart, as the recovery of the axon would permit, would stimulate B, because the first one over X_2 and the second one over X_1 would arrive at the same time. Temporal summation between two volleys would then continue as long as any branch circuits from X_3 (not shown in the diagram) are still discharging on B, when X_1 from the second volley arrives. When a strong

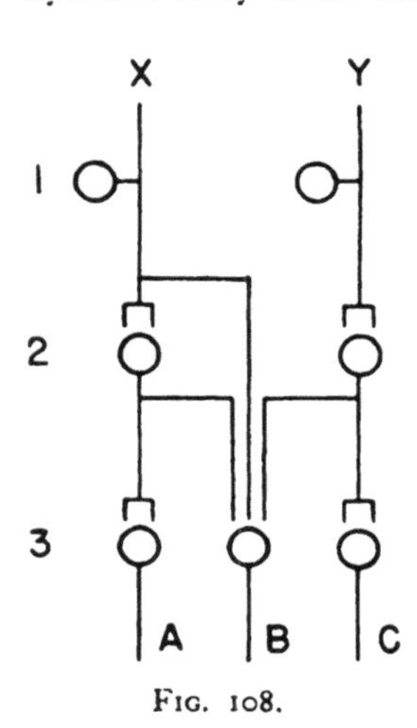

Fig. 108.

shock adds Y to X, the line is clear to motor neurone C; Y_2 sums with X_2 at B, and all three motor neurones respond.

If now we introduce the condition that the X line is stimulated, and then that after too long an interval for the Y line to be facilitated, both X and Y are excited by a stronger shock, it should follow that, the threshold of X_2 being still raised on account of the first response, two endings from X_1 would not have an effect above threshold; X_2 would not respond, and hence no excitation would go to B from the second volley by way of X_2. X_1 will not sum with Y_2, because of the synaptic delay Y_1 to Y_2, and the Y_2 ending by itself is insufficient to produce excitation. Motor neurone C would be the only one to respond. In other words, although only one motor neurone is stimulated by a weak shock, two are made unresponsive to the strong one.

It must be admitted that the synaptology of the spinal cord is much more complex than the diagram represents and that the schema should not be taken as a complete explanation of inhibition. The schema can, however, be considered as an indication of the type of mechanism which must operate if inhibition is to be explained, using only neurone linkages and the unresponsiveness of nervous tissue which follows in the wake of activity. Durations of unresponsiveness beyond those which have just been described must be accounted for if all the phenomena in the central nervous system which are termed "inhibitions" are to be considered as manifestations of the mechanism. But there are also left to be called upon, in the interest of obtaining a satisfactory explanation of a longer duration, such factors as delay circuits, reverberating arcs, and the summated effect of repeated activity.

Subnormality and the Silent Period

Another application of the after-potentials to reflex physiology is made in connection with what is known as the "silent period." When a reflex is produced by an afferent volley, the initial discharge is followed by a period during which discharge is absent. Even the background discharge, which would otherwise come into motor fibers, is erased.

The phenomenon was shown originally in a very simple way by P. Hoffmann. He recorded with a string galvanometer the action potentials from the flexor muscles of the foot during a strong, voluntary contraction, and then, while the contraction was in progress, produced a reflex twitch by a shock applied to the peroneal nerve. The record (Fig. 109) showed first the voluntary innervation, then a large action potential attributable to stimulation of the motor nerve followed by another action potential of reflex origin. The latter in turn was followed by a period in which the voluntary innervation dropped out. P. Hoffmann and others have also found that in a myotatic reflex, in which the intercurrent stimulus is over physiologically selected fibers, the same phenomenon appears, and that the effect is not confined to the stretched muscle. Spread of excitation must, therefore, take place in the nervous system beyond the connections to the one muscle. In explanation of the phenomenon nearly everyone who has dealt with it has invoked the intervention of inhibition.

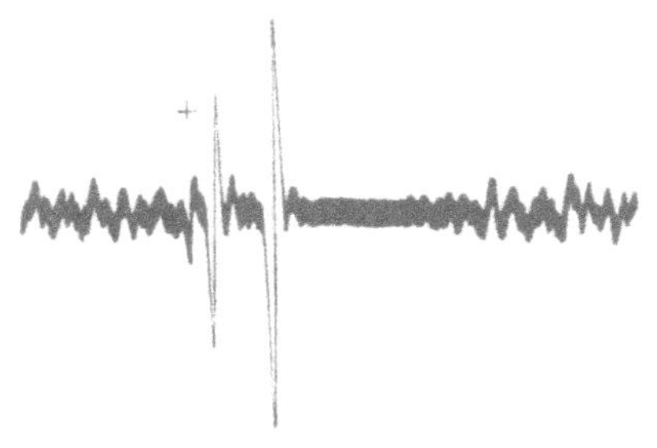

FIG. 109. Action potentials of the flexor muscles of the foot recorded with the string galvanometer. During voluntary flexion of the foot an induction shock is applied to the peroneal nerve at the time marked +. There then follows the large action potential belonging to the indirect twitch, and later that belonging to the reflex response. Following the latter the voluntary innervation is suspended. (P. Hoffmann, 1920.)

An amusing imitation of the silent period can be produced in a nerve preparation. Freshly isolated phrenic nerves, as was shown by Adrian (1930), have a marked tendency to discharge from the injured ends. One fiber after another becomes active, keeping up a continuous background of excitation. If to such a nerve a maximal shock be applied (Fig. 110), all the fibers are made active at once, and their subsequent ability to pick up excitation from the background source becomes dependent upon the excitability cycle through which they pass. Following the spike and the refractory period there is an exaggerated supernormal period, out of keeping with the conditions in the body, and it may be disregarded. The point of interest is found in the subnormal period. During that interval, the background excitation is suppressed, and it will be noted that the period of its suppression is comparable in duration to that of the silent period.

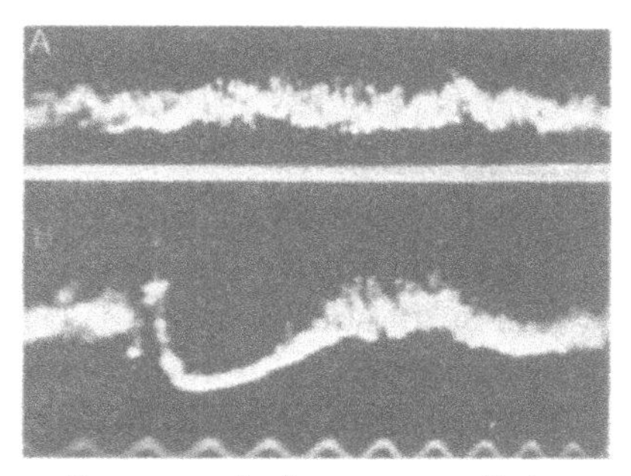

FIG. 110. A. Spontaneous discharges in an isolated phrenic nerve of the cat. The nerve was at 37° C. and in an atmosphere of oxygen. B. Changes in the discharge occurring during the response to an induction shock. The response starts at the break in the line. The spike is not visible and the first upward deflection is the negative after-potential. After 10 msec. the positive after-potential starts and lasts until 60 msec., at which time it is followed by the second negative after-potential. Time, 20 msec.

Following the subnormality, the background discharge again rises to a magnitude definitely greater than it possesses during its undisturbed, steady state. This is nothing more than the expression of the effect of the second period of supernormality, but it now has a unique interest. If we turn to the records of P. Hoffmann and of Lindsley (Fig. 111), we note that the silent period in the cord discharge is followed at a similar interval by an increased output of impulses to the muscles. Shall we associate the two? If the association held, we should not only have an explanation of a happening which in the language of the

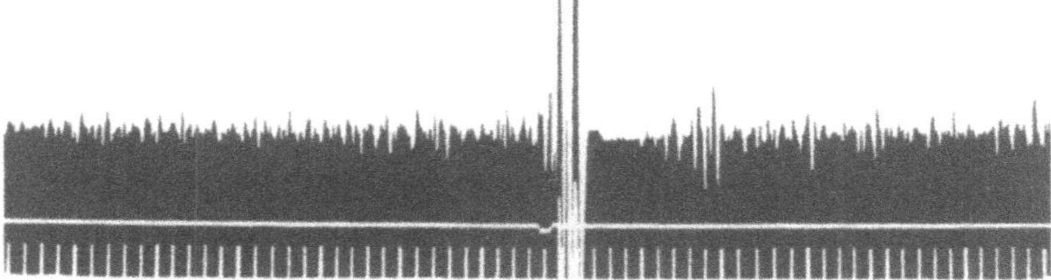

FIG. 111. Action potentials taken with gross electrodes from the rectus femoris muscle of the human subject during voluntary extension, showing a silent period following a knee-jerk. The time of the tendon tap which elicited the jerk is indicated by the signal line. Following the silent period there is an increase in the voluntary innervation. Time, 20 msec. (D. B. Lindsley, 1934.)

nervous system is termed a "rebound," but also an instance of supernormality in central neurones. At the present time it is probably not advisable to consider the similarity as being more than suggestive.

A more refined method of performing experiments of the type just mentioned is to record the discharge into single muscle units (Eccles and Hoff; Lindsley; and Hoff, Hoff, and Sheehan). The rhythm of the innervation depends upon the intensity of the background. If the latter is weak, the individual action potentials may be one-tenth of a second or more apart, thus giving ample opportunity for ascertaining the effect of interpolating a discharge between any two of them. Whether an interpolated action is aroused antidromically or reflexly, the result is the same. The added action is followed by a pause,

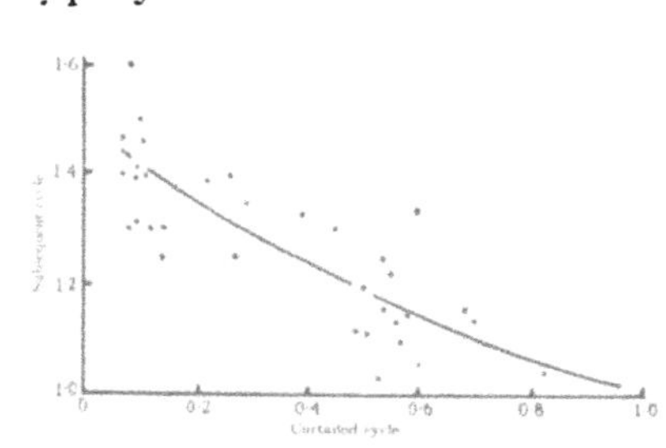

FIG. 112. The effect of single reflex discharges on the rhythm of a soleus motor neurone. Interpolated actions in rhythmically discharging motor neurones were elicited by single afferent volleys. The intervals between the interpolated actions and the last members of the regular series (curtailed cycle) are plotted as abscissae against the durations of the intervals to the next member of the regular series (subsequent cycle), as ordinates. Both intervals are expressed as fractions of the average normal cycle. (Hoff, Hoff, and Sheehan, 1934.)

the duration of which depends upon the proximity of the action to the last preceding impulse of the background series; the closer the two are together, the longer is the pause (Fig. 112).

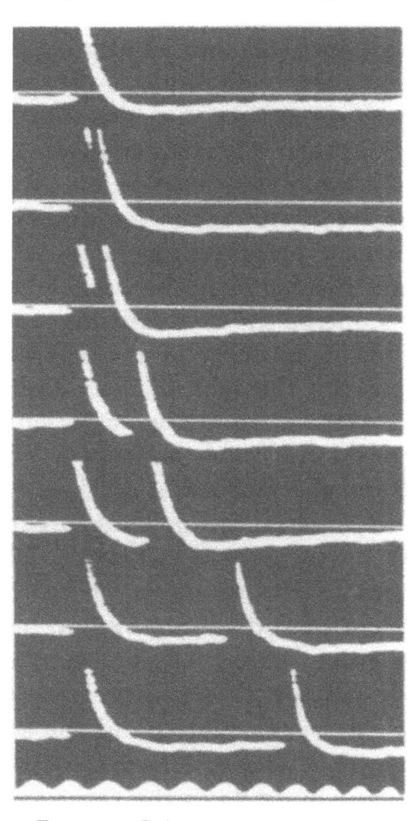

FIG. 113. Paired responses in a phrenic nerve recorded with D. C. amplification in a way so as to show the behavior of the positive after-potential. The uppermost curve shows an isolated response and the curves below it show two responses at progressively increasing intervals. It is evident that the greatest area of positive potential is produced when the responses are closest together. The potential record was made monophasic by cocaine on the distal lead. Conduction distance, 2.0 cm. Temp., 37° C. Time, 10 msec.

Since an interpolated antidromic volley has the same effect as reflex excitation, a sufficient mechanism for the production of the phenomenon must exist in the motor neurone. What this mechanism is becomes readily apparent on examination of the positive after-potential following double responses of peripheral nerve (Fig. 113). The closer are the responses, the deeper and longer is the ensuing positivity, and therefore the greater is the subnormality and the longer is the time which must elapse before a shock at constant near-threshold strength will become effective again.

When translated to the interpretation of spinal cord activity, the nerve experiments have the following significance. Active endings of various origins provide excitation, which keeps the motor neurones in a steady state of discharge. Each action in the discharge is followed by the usual subnormal period, and the latter has a determining influence on the rhythm. A faster rhythm necessitates more active endings, so that stronger stimulation may bring activity earlier into the subnormal period. Greater activity in turn increases subnormality and opposes the increased rhythm. The end result thus is derived from a balance between the two. An added single action is the simplest case of increased activity. When it comes closest to the last discharge, it has the greatest effect upon the positive potential of the neurone and the greatest capacity for producing delay of the following discharge.

From the fact that an added action in a motor neurone creates a condition sufficient for the silent period, two conclusions may be drawn. The first conclusion is derived from the observation that excitation of the neurone by impulses in the normal direction produces the same effect as antidromic excitation. The influence of unresponsiveness of the internuncial neurones on the duration of the silent period cannot extend beyond the compass of the subnormal period of the motor neurones themselves. The second conclusion is derived from the similarity in duration of the silent period and the subnormal period of the motor axons. The cell bodies of the motor neurones and their axons must have the same subnormal period.

The two simple events which have just been analyzed in terms of the properties of nervous tissue lead one to believe that other events of a more complex nature may be treated in a similar manner. The features of the central ganglionic apparatus which endow it with its qualities as an organ of integration are elaborateness of structure and plasticity of the connections between its parts. Every central reaction involves a detailed pattern of activity within the structure, and consequently an understanding of the reaction must be based on detailed information. The larger and more complex the functional event which may be examined, the greater is the possibility of picking up significant details about the operation of the mechanism. Investigation, however, will be slow and laborious; and it can proceed in an orderly manner only if it is founded on a thoroughgoing knowledge of the properties of the stuff out of which the structure is built and their relation to the ease of transmission between the units of the structure. Our knowledge of patterns can never transcend our understanding of a single synapse.

REFERENCES

Adrian, E. D. The recovery process of excitable tissues. Part II. J. Physiol. **55**, 193, 1921.

Adrian, E. D. The impulses produced by sensory nerve endings. Part I. J. Physiol. **61**, 49, 1926.

Adrian, E. D. The effects of injury on mammalian fibers. Proc. Roy. Soc. Ser. B. **106**, 596, 1930.

Adrian, E. D. The mechanism of nervous action. Philadelphia. Univ. Penn. Pr. 1932.

Adrian, E. D. Recent work on the sensory mechanism of the nervous system. Advances in Modern Biology (XV Internat. Cong. of Physiol.), Moscow, **IV**, 11, 1935.

Adrian, E. D. and Bronk, D. W. The discharge of impulses in motor nerve fibres. Part I. Impulses in single fibres of the phrenic nerve. J. Physiol. **66**, 81, 1928.

Adrian, E. D. and Bronk, D. W. The discharge of impulses in motor nerve fibres. Part II. The frequency of discharge in reflex and voluntary contractions. J. Physiol. **67**, 119, 1929.

Adrian, E. D., Cattell, McK. and Hoagland, H. Sensory discharges in single cutaneous nerve fibres. J. Physiol. **72**, 377, 1931.

Adrian, E. D. and Lucas, K. On the summation of propagated disturbances in nerve and muscle. J. Physiol. **46**, 68, 1912.

Amberson, W. R. and Downing, A. C. The electric response of nerve to two stimuli. J. Physiol. **68**, 1, 1929/1930.

Amberson, W. R. and Downing, A. C. On the form of the action potential wave in nerve. J. Physiol. **68**, 19, 1929/1930.

Amberson, W. R., Parpart, A. and Sanders, G. An analysis of the low-voltage elements of the action-potential wave in nerve. Am. J. Physiol. **97**, 154, 1931.

Beutner, R. Die Entstehung elektrischer Ströme in lebenden Geweben und ihre künstliche Nachahmung durch synthetische organische Substanzen. Ferdinand Enke, Stuttgart, 1920.

Bishop, G. H. The form of the record of the action potential of vertebrate nerve at the stimulated region. Am. J. Physiol. **82**, 462, 1927.

Bishop, G. H. The relation between the threshold of nerve response and polarization by galvanic current stimuli. Am. J. Physiol. **84**, 417, 1928.

Bishop, G. H. The effect of nerve reactance on the threshold of nerve during galvanic current flow. Am. J. Physiol. **85**, 417, 1928.

Bishop, G. H. The reactance of nerve and the effect upon it of electrical currents. Am. J. Physiol. **89**, 618, 1929.

Bishop, G. H. Action of nerve depressants on potential. J. Cell. and Comp. Physiol. **1**, 175, 1932*a*.

Bishop, G. H. The relation of nerve polarization to monophasicity of its response. J. Cell. and Comp. Physiol. **1**, 371, 1932*b*.

Bishop, G. H. Fiber groups in the optic nerve. Am. J. Physiol. **106**, 460, 1933.

Bishop, G. H. The action potentials at normal and depressed regions of non-myelinated fibers, with special reference to the "monophasic" lead. J. Cell. and Comp. Physiol. **5**, 151, 1934/1935.

Bishop, G. H., Erlanger, J. and Gasser, H. S. Distortion of action potentials as recorded from the nerve surface. Am. J. Physiol. **78**, 592, 1926.

Bishop, G. H. and Heinbecker, P. Differentiation of axon types in visceral nerves by means of the potential record. Am. J. Physiol. **94**, 170, 1930.

Bishop, G. H., Heinbecker, P. and O'Leary, J. L. Nerve degeneration accompanying experimental poliomyelitis. II. A histologic and functional analysis of normal somatic and autonomic nerves of the monkey. Arch. Neurol. and Psychiat. **27**, 1070, 1932.

Bishop, G. H., Heinbecker, P. and O'Leary, J. L. The function of the non-myelinated fibers of the dorsal roots. Am. J. Physiol. **106**, 647, 1933.

Blair, E. A. and Erlanger, J. On the effects of polarization of nerve fibers by extrinsic action potentials. Am. J. Physiol. **101**, 559, 1932.

Blair, E. A. and Erlanger, J. A comparison of the characteristics of axons through their individual electrical responses. Am. J. Physiol. **106**, 524, 1933.

Blair, E. A. and Erlanger, J. On the process of excitation by brief shocks in axons. Am. J. Physiol. **114**, 309, 1935/1936*a*.

Blair, E. A. and Erlanger, J. On excitation and depression in axons at the cathode of the constant current. Am. J. Physiol. **114**, 317, 1935/1936*b*.

Blinks, L. R. The effects of current flow on bio-electric potential. III. Nitella. J. Gen. Physiol. **20**, 229, 1936/1937.

Bogue, J. Y. and Rosenberg, H. The rate of development and spread of electrotonus. J. Physiol. **82**, 353, 1934.

Bozler, E. Untersuchungen über das Nervensystem der Coelenteraten. II. Teil: Über die Struktur der Ganglienzellen und die Funktion der Neurofibrillen nach Lebenduntersuchungen. Z. vergleich. Physiol. **6**, 255, 1927.

Bremer, F. and Titeca, J. Étude potentiométrique de la paralysie thermique du nerf. Compt. rend. Soc. biol. **115**, 413, 1933.

Bronk, D. W. The mechanism of sensory end organs. In Sensation: Its Mechanisms and Disturbances. Ass'n. for Research in Nervous and Mental Disease, **15**, Baltimore, Williams and Wilkins, 60, 1935.

Bronk, D. W. and Stella, G. The response to steady pressures of single end organs in the isolated carotid sinus. Am. J. Physiol. **110**, 708, 1934/1935.

Cajal, S., Ramón y. Histologie du Système Nerveaux. **I**, 531, 1909. Maloine, Paris.

Clark, D., Hughes, J. and Gasser, H. S. Afferent function in the group of nerve fibers of slowest conduction velocity. Am. J. Physiol. **114**, 69, 1935/1936.

Davenport, H. A. and Ranson, S. W. Ratios of cells to fibers and of myelinated to unmyelinated fibers in spinal nerve roots. Am. J. Anat. **49**, 193, 1931/1932.

de Rényi, G. S. The structure of cells in tissues as revealed by microdissection. II. The physical properties of the living axis cylinder in the myelinated nerve fiber of the frog. J. Comp. Neurol. **47**, 405, 1928/1929. III. Observations on the sheaths of myelinated nerve fibers of the frog. J. Comp. Neurol. **48**, 293, 1929.

Denny-Brown, D. On inhibition as a reflex accompaniment of the tendon jerk and of other forms of active muscular response. Proc. Roy. Soc. Ser. B. **103**, 321, 1928.

Dittler, R. Die "Reizzeit" von Induktionsschlägen verschiedener Stärke. Ztschr. f. Biol. **83**, 29, 1925.

Donaldson, H. H. and Hoke, G. W. On the areas of the axis cylinder and medullary sheath as seen in cross sections of the spinal nerves of vertebrates. J. Comp. Neurol. and Psychol. **15**, 1, 1905.

Douglass, T. C., Davenport, H. A., Heinbecker, P. and Bishop, G. H. Vertebrate nerves: some correlations between fiber size and action potentials. Am. J. Physiol. **110**, 165, 1934/1935.

Duncan, D. A relation between axone diameter and myelination determined by measurement of myelinated spinal root fibers. J. Comp. Neurol. **60**, 437, 1934.

Dunn, E. H. A statistical study of the medullated nerve fibers innervating the legs of the leopard frog, Rana pipiens, after unilateral section of the ventral roots. J. Comp. Neurol. and Psychol. **19**, 685, 1909.

Ebbecke, U. Über elektrotonische Ströme an Kollodiummembranen. Ztschr. f. Biol. **91**, 247, 1930/1931.

Eccles, J. C. Studies on the flexor reflex. III. The central effects produced by an antidromic volley. Proc. Roy. Soc. Ser. B. **107**, 557, 1930/1931.

Eccles, J. C. Slow potential waves in the superior cervical ganglion. J. Physiol. **85**, 464, 1935.

Eccles, J. C. and Hoff, H. E. The rhythmic discharge of motoneurones. Proc. Roy. Soc. Ser. B. **110**, 483, 1932.

Eccles, J. C. and Sherrington, C. S. Numbers and contraction-values of individual motor-units examined in some muscles of the limb. Proc. Roy. Soc. Ser. B. **106**, 326, 1930*a*.

Eccles, J. C. and Sherrington, C. S. Reflex summation in the ipsilateral spinal flexion reflex. J. Physiol. **69**, 1, 1930*b*.

Eccles, J. C. and Sherrington, C. S. Studies on the flexor reflex. II. The reflex response evoked by two centripetal volleys. Proc. Roy. Soc. Ser. B. **107**, 535, 1930/1931.

Erlanger, J. The interpretation of the action potential in cutaneous and muscle nerves. Am. J. Physiol. **82**, 644, 1927.

Erlanger, J. and Blair, E. A. The irritability changes in nerve in response to subthreshold induction shocks, and related phenomena, including the relatively refractory phase. Am. J. Physiol. **99**, 108, 1931/1932*a*.

Erlanger, J. and Blair, E. A. The irritability changes in nerve in response to subthreshold constant currents, and related phenomena. Am. J. Physiol. **99**, 129, 1931/1932*b*.

Erlanger, J. and Blair, E. A. Manifestations of segmentation in myelinated axons. Am. J. Physiol. **110**, 287, 1934/1935.

Erlanger, J. and Blair, E. A. Observations on repetitive responses in axons. Am. J. Physiol. **114**, 328, 1935/1936.

Erlanger, J. and Gasser, H. S. The compound nature of the action current of nerve as disclosed by the cathode ray oscillograph. Am. J. Physiol. 70, 624, 1924.

Erlanger, J. and Gasser, H. S. The action potential in fibers of slow conduction in spinal roots and somatic nerves. Am. J. Physiol. **92**, 43, 1930.

Erlanger, J., Gasser, H. S. and Bishop, G. H. The absolutely refractory phase of the alpha, beta and gamma fibers in the sciatic nerve of the frog. Am. J. Physiol. **81**, 473, 1927.

Fenn, W. O., Cobb, D. M., Hegnauer, A. H. and Marsh, B. S. Electrolytes in nerve. Am. J. Physiol. **110**, 74, 1934/1935.

Forbes, A., Davis, H. and Emerson, J. H. An amplifier, string galvanometer and photographic camera designed for the study of action currents in nerve. Rev. Sci. Instr. **2**, 1, 1931.

Furusawa, K. The depolarization of crustacean nerve by stimulation or oxygen want. J. Physiol. **67**, 325, 1929.

Garten, S. Ein Beitrag zur Kenntnis der positiven Nachschwankung des Nervenstromes nach elektrischer Reizung. Arch. ges. Physiol. (Pflüger), **136**, 545, 1910.

Gaskell, W. H. On the structure, distribution and function of the nerves which innervate the visceral and vascular systems. J. Physiol. **7**, 1, 1885–86.

Gasser, H. S. Nerve activity as modified by temperature changes. Am. J. Physiol. **97**, 254, 1931.

Gasser, H. S. Conduction in nerves in relation to fiber types. In Sensation: Its Mechanisms and Disturbances. Ass'n. for Research in Nervous and Mental Disease, **15**, Baltimore, Williams and Wilkins, 35, 1935*a*.

Gasser, H. S. The changes produced in the injury discharge of nerve fibers during the period following a propagated disturbance. J. Physiol. **85**, 15 P, 1935*b*.

Gasser, H. S. Changes in nerve-potentials produced by rapidly repeated stimuli and their relation to the responsiveness of nerve to stimulation. Am. J. Physiol. **111**, 35, 1935*c*.

Gasser, H. S. and Erlanger, J. The nature of conduction of an impulse in the relatively refractory period. Am. J. Physiol. **73**, 613, 1925.

Gasser, H. S. and Erlanger, J. The rôle played by the sizes of the constituent fibers of a nerve trunk in determining the form of its action potential wave. Am. J. Physiol. **80**, 522, 1927.

Gasser, H. S. and Erlanger, J. The ending of the axon action potential, and its relation to other events in nerve activity. Am. J. Physiol. **94**, 247, 1930.

Gasser, H. S. and Graham, H. T. The end of the spike potential of nerve and its relation to the beginning of the after-potential. Am. J. Physiol. **101**, 316, 1932.

Gasser, H. S. and Graham, H. T. Potentials produced in the spinal cord by stimulation of dorsal roots. Am. J. Physiol. **103**, 303, 1933.

Gasser, H. S. and Grundfest, H. Action and excitability in mammalian A fibers. Am. J. Physiol. **117**, 113, 1936.

Gasser, H. S. and Newcomer, H. S. Physiological action currents in the phrenic nerve. An application of the thermionic vacuum tube to nerve physiology. Am. J. Physiol. **57**, 1, 1921.

Gerard, R. W. The response of nerve to oxygen lack. Am. J. Physiol. **92**, 498, 1930.

Gerard, R. W. Delayed action potentials in nerve. Am. J. Physiol. **93**, 337, 1930.

Göthlin, G. F. Experimentella undersökninger af ledningens natur i den hvita nervsubstansen. Akad. Afhandling. Uppsala, Almquist and Wiksell, p. 57, 1907.

Graham, H. T. Modification of the response of nerve by veratrine and by narcotics. J. Pharm. and Exp. Therap. **39**, 268, 1930.

Graham, H. T. Modification of nerve after-potential and refractory period by changes of ionic environment: new cases of physiological antagonism between univalent and bivalent cations. Am. J. Physiol. **104**, 216, 1933.

Graham, H. T. Supernormality, a modification of the recovery process in nerve. Am. J. Physiol. **110**, 225, 1934/1935.

Graham, H. T. The subnormal period of nerve response. Am. J. Physiol. **111**, 452, 1935*a*.

Graham, H. T. Agents modifying the effect of subthreshold induction shocks. Am. J. Physiol. **113**, 52, 1935*b*.

Graham, H. T. and Gasser, H. S. Modification of nerve response by veratrine, protoveratrine and aconitine. J. Pharm. and Exp. Therap. **43**, 163, 1931.

Graham, H. T. and Gasser, H. S. Augmentation of the positive after-potential of nerves by yohimbine. Proc. Soc. Exp. Biol. and Med. **32**, 553, 1934/1935.

Granit, R. The components of the retinal action potential in mammals and their relation to the discharge in the optic nerve. J. Physiol. **77**, 207, 1932/1933.

Harris, A. S. A study in reflexes: identification of the cutaneous afferent fibers which evoke ipsilateral extensor and flexor reflexes. Am. J. Physiol. **112**, 231, 1935.

Hartline, H. K. A quantitative and descriptive study of the electric response to illumination of the arthropod eye. Am. J. Physiol. **83**, 466, 1927/1928.

Hartline, H. K. and Graham, C. H. Nerve impulses from single receptors in the eye. J. Cell. and Comp. Physiol. **1**, 277, 1932.

Hatai, S. On the length of the internodes in the sciatic nerve of Rana temporaria (fusca) and Rana pipiens: being a reexamination by biometric methods of the data studied by Boycott ('04) and Takahashi ('08). J. Comp. Neurol. and Psychol. **20**, 19, 1910.

Heinbecker, P. Properties of unmyelinated fibers of nerve. Proc. Soc. Exp. Biol. and Med. **26**, 349, 1928/1929.

Heinbecker, P. The effect of fibers of specific types in the vagus and sympathetic nerves on the sinus and atrium of the turtle and frog heart. Am. J. Physiol. **98**, 220, 1931.

Heinbecker, P. and Bishop, G. H. Studies on the extrinsic and intrinsic nerve mechanisms of the heart. Am. J. Physiol. **114**, 212, 1935/1936.

Heinbecker, P., Bishop, G. H. and O'Leary, J. Analysis of sensation in terms of the nerve impulse. Arch. Neurol. and Psychiat. **31**, 34, 1934.

Heinbecker, P., O'Leary, J. and Bishop, G. H. Nature and source of fibers contributing to the saphenous nerve of the cat. Am. J. Physiol. **104**, 23, 1933.

Hermann, L. Zur Physiologie und Physik des Nerven. Arch. ges. Physiol. (Pflüger) **109**, 95, 1905.

Hill, A. V. Chemical wave transmission in nerve. London, 1932, p. 27.

Hill, A. V. The three phases of nerve heat production. Proc. Roy. Soc. Ser. B. **113**, 345, 1933.

Hill, A. V. The effect of veratrine on the heat production of medullated nerve. Proc. Roy. Soc. Ser. B. **113**, 386, 1933.

Hill, A. V. Excitation and accommodation in nerve. Proc. Roy. Soc. Ser. B. **119**, 305, 1935/1936.

Höber, R. and Strohe, H. Über den Einfluss von Salzen auf die elektrotonischen Ströme, die Erregbarkeit und das Ruhepotential des Nerven. Arch. ges. Physiol. (Pflüger) **222**, 71, 1929.

Hoff, E. C. Central nerve terminals in the mammalian spinal cord and their examination by experimental degeneration. Proc. Roy. Soc. Ser. B. **111**, 175, 1932.

Hoff, E. C., Hoff, H. E. and Sheehan, D. Reflex interruptions of rhythmic discharge. J. Physiol. **83**, 185, 1934/1935.

Hoffmann, P. Demonstration eines Hemmungsreflexes im menschlichen Rückenmark. Z. f. Biol. **70**, 515, 1919/1920.

Hughes, J. and Gasser, H. S. Some properties of the cord potentials evoked by a single afferent volley. Am. J. Physiol. **108**, 295, 1934*a*.

Hughes, J. and Gasser, H. S. The response of the spinal cord to two afferent volleys. Am. J. Physiol. **108**, 307, 1934*b*.

Ide, K. On the changes in the sectional areas of the largest fibers in the fifth lumbar spinal nerve and its roots in the albino rat. J. Comp. Neurol. **53**, 479, 1931.

Kato, G. et al. Explanation of Wedensky Inhibition. Part II. Explanation of "Paradoxes Stadium" in the sense of Wedensky. Am. J. Physiol. **89**, 692, 1929.

Labes, R. and Zain, H. Ein Membranmodell für eine Reihe bioelektrischer Vorgänge. IV. Mitteilung: Nachahmung katelektrotonischer und anelektrotonischer Erscheinungen mittels einer Kollodiumhülsenanordnung. Arch. exp. Path. u. Pharm. **126**, 352, 1927.

Lapicque, L. Excitation par double condensateur. Compt. rend. Soc. Biol. **64**, 336, 1908.

Lapicque, L. L'excitabilité en fonction du temps; la chronaxie, sa signification et sa mesure. Paris, Les Presses Universitaires de France, 1926.

Lapicque, L. La chronaxie en biologie générale. Biol. Rev. **10**, 483, 1935.

Lapicque, L. and Legendre, R. Relation entre le diamètre des fibres nerveuses et leur rapidité fonctionelle. Compt. rend. Acad. Sci. **157**, 1163, 1913.

Lehmann, J. E. Effect of changes in the ionic constitution of the medium upon the after-potentials of mammalian fibers. Unpublished observations. 1936.

Levin, A. Fatigue, retention of action current and recovery in crustacean nerve. J. Physiol. **63**, 113, 1927.

Lillie, R. S. Factors affecting transmission and recovery in the passive iron nerve model. J. Gen. Physiol. **7**, 473, 1924/1925.

Lindsley, D. B. Inhibition as an accompaniment of the knee jerk. Am. J. Physiol. **109**, 181, 1934.

Lorente de Nó, R. The refractory period of motoneurones. Am. J. Physiol. **111**, 283, 1935*a*.

Lorente de Nó, R. The effect of an antidromic impulse on the response of the motoneurone. Am. J. Physiol. **112**, 595, 1935*b*.

Lorente de Nó, R. Facilitation of motoneurones. Am. J. Physiol. **113**, 505, 1935*c*.

Lorente de Nó, R. The summation of impulses transmitted to the motoneurones through different synapses. Am. J. Physiol. **113**, 524, 1935*d*.

Lorente de Nó, R. and Graham, H. T. Recovery of mammalian nerve fibers *in vivo*. Proc. Soc. Exp. Biol. and Med. **33**, 512, 1935/1936.
Lucas, K. Quantitative researches on the summation of inadequate stimuli in muscle and nerve, with observations on the time factor in electric excitation. J. Physiol. **39**, 461, 1909/1910.
Lullies, H. Über die Polarisation in Geweben. III. Mit. Die Polarisation im Nerven II. Arch. ges. Physiol. (Pflüger) **225**, 87, 1930.
Marshall, W. H. and Gerard, R. W. Nerve impulse velocity and fiber diameter. Am. J. Physiol. **104**, 586, 1933.
Matthews, B. H. C. A new electrical recording system for physiological work. J. Physiol. **65**, 225, 1928.
Matthews, B. H. C. Nerve endings in mammalian muscle. J. Physiol. **78**, 1, 1933.
Michaelis, L. Die getrocknete Kollodiummembran. Koll. Z. **62**, 2, 1933.
Michaelis, L. and Fujita, A. Untersuchungen über elektrische Erscheinungen und Ionendurchlässigkeit von Membranen. IV. Mit. Potentialdifferenzen und Permeabilität von Kollodiummembranen. Biochem. Ztschr. **161**, 47, 1925.
Monnier, A. M. L'excitation electrique des tissus. Paris, Hermann, 1934.
O'Leary, J., Heinbecker, P. and Bishop, G. H. The fiber constitution of the depressor nerve of the rabbit. Am. J. Physiol. **109**, 274, 1934.
O'Leary, J., Heinbecker, P. and Bishop, G. H. Analysis of function of a nerve to muscle. Am. J. Physiol. **110**, 636, 1934/1935.
Parker, G. H. The neurofibril hypothesis. Quart. Rev. Biol. **4**, 155, 1929.
Ranson, S. W. and Billingsley, P. R. The conduction of painful afferent impulses in the spinal nerves. Studies in vasomotor reflex arcs. II. Am. J. Physiol. **40**, 571, 1916.
Ranson, S. W. and Hinsey, J. C. Reflexes in the hind limbs of cats after transection of the spinal cord at various levels. Am. J. Physiol. **94**, 471, 1930.

Richards, C. H. and Gasser, H. S. After-potentials and recovery curve of C fibers. Am. J. Physiol. **113**, 108, 1935.

Schmitt, F. O. and Gasser, H. S. The relation between the after-potential and oxidative processes in medullated nerve. Am. J. Physiol. **104**, 320, 1933.

Schmitz, W. and Schaefer, H. Ladekurve, Ladezeit und Latenzzeit der Aktion bei elektrischer Nervenreizung. Arch. ges. Physiol. (Pflüger), **233**, 229, 1933/1934.

Sherrington, C. S. Quantitative management of contraction in lowest level co-ordination. Brain **54**, 1, 1931.

Tigerstedt, R. Zur Theorie der Oeffnungszuckung. Stockholm, Norstedt, 1882. Suppl. to K. Svenska vet.-acad. handl. VII, No. 7, 1882.

Verzár, F. Über die Natur der Thermoströme des Nerven. Arch. ges. Physiol. (Pflüger) **143**, 252, 1912.

Winkler, C. Manuel de Neurologie, Haarlem, Bohn, 1929, Vol. **1**, 189.

Wyss, O. A. M. Das Problem der selektiven elektrischen Nervenreizung. Schweiz. Arch. Neurol. and Psychiat. **28**, 210, 1931/1932.

Young, J. Z. Structure of nerve fibers in Sepia. J. Physiol. **83**, 27 P, 1934/1935.

BIBLIOGRAPHY OF JOSEPH ERLANGER
1874–1965

1. ERLANGER, J., and A. W. HEWLETT. 1901. A study of the metabolism in dogs with shortened small intestines. *Amer. J. Physiol.* **6**: 1-30.
2. ERLANGER, J. 1903-04. A study of the errors involved in the determination of the blood-pressures in man together with a demonstration of the improvements in the sphygmomanometer thereby suggested. (Proc. Amer. Physiol. Soc.) *Amer. J. Physiol.* **10**: xiv.
3. ERLANGER, J., and D. R. HOOKER. 1903-04. The relation of blood-pressure and pulse-pressure to the secretion of urine and to the secretion of albumin in a case of so-called physiological albuminuria. (Proc. Amer. Physiol. Soc.) *Amer. J. Physiol.* **10**: xvi.
4. ERLANGER, J., and D. R. HOOKER. 1903-04. The relation between blood- and pulse-pressure, and the velocity of blood-flow in man. (Proc. Amer. Physiol. Soc.) *Amer. J. Physiol.* **10**: xv.
5. ERLANGER, J. 1904. A new instrument for determining the minimum and maximum blood-pressures in man. *Johns Hopkins Hosp. Rep.* **12**: 53-110.
6. ERLANGER, J., and D. R. HOOKER. 1904. An experimental study of blood-pressure and pulse-pressure in man. *Johns Hopkins Hosp. Rep.* **12**: 145-378.
7. ERLANGER, J., and D. R. HOOKER. 1904. Studies in blood-pressure. (Abstr.) *Johns Hopkins Hosp. Bull.* **15**: 179.
8. ERLANGER, J. 1905. Cardiograms obtained from a case of operative defect in the chest wall. *Johns Hopkins Hosp. Bull.* **16**: 394-397.
9. ERLANGER, J. 1905-06. On the physiology of heart-block in mammals, with special reference to the causation of Stokes-Adams disease. *J. Exp. Med.* **7**: 676-724. **8**: 8-56.
10. ERLANGER, J. 1905. On the union of a spinal nerve with the vagus nerve. *Amer. J. Physiol.* **13**: 372-395.
11. ERLANGER, J. 1905. A report of some observations on heart-block in mammals. (Abstr.) *Johns Hopkins Hosp. Bull.* **16**: 234.

12. Erlanger, J. 1905-06. Vorlaeufige Mitterlung ueber die Physiologie des Herzblocks in Saugetieren. *Zentralbl. Physiol.* **19**: 9-12.
13. Erlanger, J., and A. D. Hirschfelder. 1905. Eine vorlaeufige Mitteilung ueber weitere Studien in bezug auf den Herzblock in Saugetieren. *Zenbl. Physiol.* **19**: 270-275.
14. Erlanger, J., and A. D. Hirschfelder. 1905-06. Further studies on the physiology of heart-block in mammals. *Amer. J. Physiol.* **15**: 153-206.
15. Erlanger, J. 1906. The effects of extra stimuli upon the heart, in the several stages of block, together with a theory of heart-block. (Abstr.) *Amer. Med.* n.s. **1**: 37.
16. Erlanger, J. 1906. Further studies on the physiology of heart-block. The effects of extra systoles upon the dog's heart and upon strips of terrapin's ventricle in the various stages of block. *Amer. J. Physiol.* **16**: 160-187.
17. Erlanger, J. 1906. Recent contributions to the physiology of the circulation. Read in the Section on Practice of Medicine of the American Medical Association, at the 57th Annual Session, June 1906. *J.A.M.A.* **47**: 1343-1351.
18. Erlanger, J. 1906. A review of the physiology of heart-block in mammals. *Brit. Med. J.* **2**: 1111-1113.
19. Erlanger, J., and T. S. Cullen. 1906. Experimental heart-block. *Johns Hopkins Hosp. Bull.* **17**: 234.
20. Erlanger, J., and J. R. Blackman. 1907. A study of relative rhythmicity and conductivity in various regions of the auricles of the mammalian heart. *Amer. J. Physiol.* **19**: 125-174.
21. Erlanger, J., J. R. Blackman, and E. K. Cullen. 1907-08. Further studies in the physiology of heart-block in mammals. Chronic auriculo-ventricular heart-block in the dog. (Proc. Amer. Physiol. Soc.) *Amer. J. Physiol.* **21**: xviii.
22. Erlanger, J. 1908. Irregularities of the heart resulting from distributed conductivity. *Amer. J. Med. Sci.* n.s. **135**: 797-811.
23. Erlanger, J. 1908-09. A method of studying the physiology of mammalian heart tissue. (Proc. Amer. Physiol. Soc.) *Amer. J. Physiol.* **23**: xxxiii-xxxvii.
24. Erlanger, J. 1908. A new criterion for the determination of the systolic blood pressure with the sphygmomanometer (with demonstration). (Proc. Amer. Physiol. Soc.) *Amer. J. Physiol.* **21**: xxiv-xxv.

25. ERLANGER, J. 1909. Can functional union be re-established between the mammalian auricles and ventricles after destruction of a segment of the auriculo-ventricular bundle? *Amer. J. Physiol.* **24**: 375-383.
26. ERLANGER, J. 1909-10. Chronic auriculo-ventricular heart-block in the dog. *Wisconsin Med. J.* **8**: 624-631.
27. ERLANGER, J. 1909-10. Mammalian heart strips together with a theory of cardiac inhibition. (Proc. Amer. Physiol. Soc.) *Amer. J. Physiol.* **25**: xvi.
28. ERLANGER, J. 1909-10. The role of the practicing physician in the defense of medical research. *Wisconsin Med. J.* **8**: 543-548.
29. ERLANGER, J. 1909. Ueber den Grad der Vaguswirkung auf die Kammern des Hundeherzens. *Arch. ges. Physiol.* **127**: 77-98.
30. ERLANGER, J., and J. R. BLACKMAN. 1909-10. Further studies in the physiology of heart-block in mammals. Chronic auriculo-ventricular heart-block in the dog. *Heart.* **1**: 177-230.
31. ERLANGER, J. 1910. Animal experimentation in relation to practical medical knowledge of the circulation. *Defense of Research.* Pamphlet #13. Council on Defense of Medical Research, American Medical Association.
Also in: *J.A.M.A.* **54**: 1680-1684, 1769-1774, 1856-1859.
32. ERLANGER, J. 1910. Observations on auricular strips of the cat's heart. *Amer. J. Physiol.* **27**: 87-119.
33. ERLANGER, J. 1910. Mammalian heart strips together with a theory of cardiac inhibition. (Proc. Amer. Physiol. Soc.) *Amer. J. Physiol.* **16**.
34. ERLANGER, J. 1912. A criticism of the Uskoff sphygotonograph. *Arch. Int. Med.* **9**: 22-31.
35. ERLANGER, J. 1912-13. The localization of impulses, initiation and conduction in the heart. (Abstr.) *New York Med. J.* **96**: 1020-1022.
Also in: *Harvey Lecture* 1912-13 ser. **8**: 44-85.
Also in: *Arch. Int. Med.* **11**: 334-364.
36. ERLANGER, J. 1912-13. Observations on the physiology of Purkinje tissue. *Amer. J. Physiol.* **30**: 395-419.
37. ERLANGER, J. 1912. Sinus stimulation as a factor in the resuscitation of the heart. *J. Exp. Med.* **16**: 452-469.

38. ERLANGER, J., and E. G. FESTERLING. 1912. Respiratory waves of blood pressure, with an investigation of a method for making continuous blood pressure records in man. *J. Exp. Med.* **15**: 370-386.
39. ERLANGER, J., W. B. CANNON, G. CRILE, et al. 1914. Report of the Committee on Resuscitation From Mine Gases. Bureau of Mines, Dept. of the Interior. *Technical Paper* #**77**. Washington: Gov. Print. Off. 36p.
40. ERLANGER, J., and W. E. GARREY. 1914. Faradic stimuli; a physical and physiological study. (Proc. Amer. Physiol. Soc.) *Amer. J. Physiol.* **35**: xiv.
Also in: *Ibid.* 377-473.
41. ERLANGER, J., and R. A. GESELL. 1914. Device for interrupting a continuous blast of air, designed especially for artificial respiration. (Proc. Amer. Physiol. Soc.) *Amer. J. Physiol.* **33**: xxxiii-xxxv.
Also in: *Ibid.* 1-10.
42. ERLANGER, J. 1915. An analysis of Dr. Kilgore's paper: "The large personal factor in blood pressure determinations by the oscillatory method." *Arch. Int. Med.* **16**: 917-926.
43. ERLANGER, J. 1915-16. Studies on blood pressure estimations by indirect methods. I. The mechanism of the oscillatory criteria. *Amer. J. Physiol.* **42**: 588-589.
44. ERLANGER, J. 1916. Blood pressure. *Amer. J. Physiol.* **39**: 401.
45. ERLANGER, J. 1916. The movements of the artery within the compression chamber during indirect estimations of the blood pressure. *Amer. J. Physiol.* **42**: 588-589.
46. ERLANGER, J. 1916. A note on the contractility of the musculature of the auriculo-ventricular valves. *Amer. J. Phsyiol.* **40**: 150-151.
47. ERLANGER, J. 1916. Studies in blood pressure estimation by indirect methods. II. The mechanism of the compression sounds of Korotkoff. *Amer. J. Physiol.* **40**: 82-125.
48. ERLANGER, J. 1917. Further observations on the mechanism of the artificial compression sounds of Korotkoff. The preanacrotic phenomenon with demonstration. *J. Missouri Med. Assoc.* **14**: 258-259.

49. ERLANGER, J. 1917. The role of Boyle's law in clinical sphygmomanometry; a reply to A. M. Bloile. *Science* n.s. **45**: 384.

50. ERLANGER, J., and R. A. GESELL. 1917. An experimental study of surgical shock; preliminary report. *J.A.M.A.* **69**: 2089-2092.

51. ERLANGER, J., and R. T. WOODYATT. 1917. Intravenous glucose injections in shock. *J.A.M.A.* **69**: 1410-1414.

52. ERLANGER, J., and H. S. GASSER. 1918. The treatment of standardized shock. I. *Compt. rend. Soc. de biol.* **81**: 898-905.

53. ERLANGER, J., and H. S. GASSER. 1918. The treatment of standardized shock. II. *Compt. rend. Soc. de biol.* **81**: 905-909.

54. ERLANGER, J., and H. S. GASSER. 1919. Hypertonic gum acacia and glucose in the treatment of secondary traumatic shock. *Ann. Surg.* **69**: 389-421.

55. ERLANGER, J., and H. S. GASSER. 1919. The treatment of standardized shock. *J. Missouri Med. Assoc.* **16**: 98.

56. ERLANGER, J., R. A. GESELL, and H. S. GASSER. 1919. Studies in secondary traumatic shock. I. The circulation in shock after abdominal injuries. *Amer. J. Physiol.* **49**: 90-116.

57. ERLANGER, J., and H. S. GASSER. 1919. Studies in secondary traumatic shock. II. Shock due to mechanical limitation of blood flow. *Amer. J. Physiol.* **49**: 151-173.

58. ERLANGER, J., and H. S. GASSER. 1919. Studies in secondary traumatic shock. III. Circulatory failure due to adrenalin. *Amer. J. Physiol.* **49**: 345-376.

59. ERLANGER, J., H. S. GASSER, and W. J. MEEK. 1919-20. Studies in secondary traumatic shock. IV. The blood volume changes and the effect of gum acacia on their development. *Amer. J. Physiol.* **50**: 31-53.

60. ERLANGER, J., and H. S. GASSER. 1919-20. Studies in secondary traumatic shock. V. Restoration of the plasma volume and of the alkali reserve. *Amer. J. Physiol.* **50**: 104-118.

61. ERLANGER, J., and H. S. GASSER. 1919-20. Studies in secondary traumatic shock. VI. Statistical study of the treatment of measured trauma with solutions of gum acacia and crystalloids. *Amer. J. Physiol.* **50**: 119-148.

62. Erlanger, J., and H. S. Gasser. 1919-20. Studies in secondary traumatic shock. VII. Note on the action of hypertonic gum acacia and glucose after hemorrhage. *Amer. J. Physiol.* **50**: 149-156.
63. Erlanger, J. 1920. Studies in blood pressure estimation by indirect methods; movements in artery under compression during blood pressure determination. *Amer. J. Physiol.* **55**: 84.
64. Erlanger, J. 1920-21. Studies in blood pressure estimation by indirect methods. III. The movements in the artery under compression during blood pressure determinations. *Amer. J. Physiol.* **55**: 158.
65. Erlanger, J., C. M. Jackson, et al. 1920. An investigation of conditions in the departments of the preclinical sciences; report of a Committee of the Division of the Medical Sciences of the National Research Council. *J.A.M.A.* **74**: 1117-1122.
66. Erlanger, J., and H. L. White. 1920-21. The effect on the composition of the blood of maintaining an increased blood volume by the intravenous injection of hypertonic solution of gum acacia and glucose in normal, asphyxiated and shocked dogs. *Amer. J. Physiol.* **54**: 1-29.
67. Erlanger, J. 1921. Blood volume and its regulation. *Physiol. Rev.* **1**: 177-207.
68. Erlanger, J., and H. L. White. 1921. Blood analysis following acacia glucose injection. *J. Missouri Med. Assoc.* **18**: 103-104.
69. Erlanger, J. 1922. The past and the future of the medical sciences in the United States. *Science* n.s. **55**: 135-145.
70. Erlanger, J., and H. S. Gasser. 1922. A study of the action currents of nerve with the cathode ray oscillograph. *Amer. J. Physiol.* **62**: 496-524.
71. Erlanger, J., and H. S. Gasser. 1924. The compound nature of the action current of nerve as disclosed by the cathode ray oscillograph. *Amer. J. Physiol.* **70**: 624-666.
72. Erlanger, J., G. H. Bishop, and H. S. Gasser. 1925. Data on the nature of propagation in nerve. *Amer. J. Physiol.* **72**: 99.
73. Erlanger, J., G. H. Bishop, and H. S. Gasser. 1925. Further observations on analysis of the action current in nerve. *Amer. J. Physiol.* **72**: 197-198.

74. Erlanger, J., and H. S. Gasser. 1925. The nature of conduction of an impulse in the relatively refractory period. *Amer. J. Physiol.* **73**: 613-635.
75. Erlanger, J., G. H. Bishop, and H. S. Gasser. 1926. The action potential waves transmitted between the sciatic nerve and its spinal roots. *Amer. J. Physiol.* **78**: 574-591.
76. Erlanger, J., G. H. Bishop, and H. S. Gasser. 1926. Distortion of action potentials as recorded from the nerve surface. *Amer. J. Physiol.* **78**: 592-609.
77. Erlanger, J., G. H. Bishop, and H. S. Gasser. 1926. The effects of polarization upon the activity of vertebrate nerve. *Amer. J. Physiol.* **78**: 630-657.
78. Erlanger, J., G. H. Bishop, and H. S. Gasser. 1926. Experimental analysis of the simple action potential wave in nerve by the cathode ray oscillograph. *Amer. J. Physiol.* **78**: 537-573.
79. Erlanger, J., G. H. Bishop, and H. S. Gasser. 1926. The record of the action potential of nerve at the site of stimulation. (Proc. Amer. Physiol. Soc.) *Amer. J. Physiol.* **76**: 204.
80. Erlanger, J., G. H. Bishop, and H. S. Gasser. 1926. The refractory phase in relation to the action potential of nerve. (Proc. Amer. Physiol. Soc.) *Amer. J. Physiol.* **76**: 203.
81. Erlanger, J., and W. J. Meek. 1926-27. An adjustable sphygmoscope for the recording sphygmomanometer. *J. Lab. Clin. Med.* **12**: 172-182.
82. Erlanger, J. 1927. The interpretation of the action potential in cutaneous and muscle nerves. *Amer. J. Physiol.* **82**: 644-655.
83. Erlanger, J., G. H. Bishop, and H. S. Gasser. 1927. The absolute refractory phase of the alpha, beta and gamma fibers in the sciatic nerve of the frog. *Amer. J. Physiol.* **81**: 473-474.
84. Erlanger, J., and H. S. Gasser. 1927. The differential action of pressure on fibers of different sizes in a mixed nerve. *Proc. Soc. Exper. Biol. & Med.* **24**: 313-314.
85. Erlanger, J., and H. S. Gasser. 1927. The role played by the sizes of the constituent fibers of a nerve trunk in determining the form of its action potential wave. *Amer. J. Physiol.* **80**: 522-547.
86. Erlanger, J. 1926-27-28. Analysis of the action potential in nerve. *Harvey Lectures*. ser. **22**: 90-113.

87. Erlanger, J., and H. S. Gasser. 1928-29-30. The action potential in fibers of slow conduction in spinal roots and somatic nerves. *Proc. Soc. Exper. Biol. & Med.* **26**: 647-649. Also in: *Amer. J. Physiol.* **92**: 43-82.

88. Erlanger, J., and F. P. Schmitt. 1928. Directional differences in conduction of impulse through heart muscle and their possible relation to extrasystolic and fibrillary contractions. *Amer. J. Physiol.* **87**: 326-347.

89. Erlanger, J., and H. S. Gasser. 1929. Role of fiber size in establishment of nerve block by pressure of cocaine. *Amer. J. Physiol.* **88**: 581-591.

90. Erlanger, J., and H. S. Gasser. 1930. Action potential in fibers of slow conduction in spinal roots and somatic nerves. *Amer. J. Physiol.* **92**: 43-82.

91. Erlanger, J., and H. S. Gasser. 1930. The ending of the axon action potential and its relation to other events in nerve activity. *Amer. J. Physiol.* **94**: 247-277.

92. Erlanger, J., and E. A. Blair. 1931-32. Irritability changes in nerve in response to subthreshold induction shocks, and related phenomena including relatively refractory phase. *Amer. J. Physiol.* **99**: 108-128.

93. Erlanger, J., and E. A. Blair. 1931-32. Irritability changes in nerve in response to subthreshold constant currents and related phenomena. *Amer. J. Physiol.* **99**: 129-155.

94. Erlanger, J., and E. A. Blair. 1932. On effects of polarization of nerve fibers by extrinsic action potentials. *Amer. J. Physiol.* **101**: 559-564.

95. Erlanger, J., and E. A. Blair. 1932. Responses of axons to brief shocks. *Proc. Soc. Exp. Biol. & Med.* **29**: 926-927.

96. Erlanger, J. 1933. William Beaumont's experiments and their present day value. *Weekly Bull. St. Louis Med. Soc.* **28**: 180-191.

97. Erlanger, J., and E. A. Blair. 1933. A comparison of the characteristics of axons through their individual electrical responses. *Amer. J. Physiol.* **106**: 524-564.

98. Erlanger, J., and E. A. Blair. 1933. Comparison of properties of individual axons in the frog. *Proc. Soc. Exp. Biol. & Med.* **30**: 728-729.

99. ERLANGER, J., and E. A. BLAIR. 1933. The configuration of axon and "simple" nerve action potentials. *Amer. J. Physiol.* **106**: 565-570.
100. ERLANGER, J., and E. A. BLAIR. 1933. Reply to Bishop and Heinbecker's "Fiber distribution in optic and saphenous nerves." *Proc. Soc. Exp. Biol. & Med.* **31**: 127-128.
101. ERLANGER, J., and E. A. BLAIR. 1934-35. Manifestation of segmentation in myelinated axons. *Amer. J. Physiol.* **100**: 287-311.
102. ERLANGER, J., and E. A. BLAIR. 1935-36. Observations on repetitive responses in axons. *Amer. J. Physiol.* **114**: 328-361.
103. ERLANGER, J., and E. A. BLAIR. 1935-36. On excitation and depression in axons at the cathode of the constant current. *Amer. J. Physiol.* **114**: 317-327.
104. ERLANGER, J., and E. A. BLAIR. 1935-36. On the process of excitation by brief shocks in axons. *Amer. J. Physiol.* **114**: 309-316.
105. ERLANGER, J., and E. A. BLAIR. 1936. Temporal summation in peripheral nerve fibers. *Amer. J. Physiol.* **117**: 355-365.
106. ERLANGER, J. 1937. Brookings: A Biography, by Hermann Hagedorn (review). *J. Missouri Med. Assoc.* **34**: 138.
107. ERLANGER, J., and H. S. GASSER. 1937. Electrical signs of nervous activity. Eldridge Reeves Johnson Foundation for Medical Physics. Philadelphia: Univ. of Penna. Press. 221p.
108. ERLANGER, J., and E. A. BLAIR. 1938. The action of isotonic, salt-free solutions on conduction in medullated nerve fibers. *Amer. J. Physiol.* **124**: 341-359.
109. ERLANGER, J., and E. A. BLAIR. 1938. Comparative observations on motor and sensory fibers with special reference to repetitiousness. *Amer. J. Physiol.* **121**: 431-453.
110. ERLANGER, J. 1939. The initiation of impulses in axons. *J. Neurophysiol.* **2**: 370-379.
111. ERLANGER, J., and E. A. BLAIR. 1939. Propagation, and extension of excitatory effects, of the nerve action potential across nonresponding internodes. *Amer. J. Physiol.* **126**: 97-108.
112. ERLANGER, J. 1940. The relation of longitudinal tension of an artery to the preanacrotic (breaker) phenomenon. *Amer. Heart J.* **19**: 398-400.

113. ERLANGER, J., and E. A. BLAIR. 1940. Facilitation and difficilitation effected by nerve impulses in peripheral fibers. *J. Neurophysiol.* 3: 107-127.
114. ERLANGER, J., and E. A. BLAIR. 1940-41. Interaction of medullated fibers of a nerve tested with electric shocks. *Amer. J. Physiol.* **131**: 483-493.
115. ERLANGER, J. 1941. Remarks on some evidence of subconducted process in medullated nerve fibers. (Issue 12 in honor of Dr. W. R. Hess.) *Schweiz. Med. Wschnchr.* **22**: 394-395.
116. ERLANGER, J. 1941. Standardization of blood pressure readings. Joint recommendations of the Amer. Heart Assoc. and Cardiac Soc. of Great Britain & Ireland. New York: *Amer. Heart Assoc.* 8p.
117. ERLANGER, J., and G. M. SCHOEPFLE. 1941. The action of temperature on the excitability, spike height and configuration, and the refractory period observed in the responses of single medullated nerve fibers. *Amer. J. Physiol.* **134**: 694-704.
118. ERLANGER, J., A. G. KREMS, and G. M. SCHOEPFLE. 1942. Nerve concussion. *Proc. Soc. Exp. Biol. & Med.* **49**: 73-75.
119. ERLANGER, J. 1943. *Correspondence. Amer. Heart J.*, **26**: 419-420. Letter to the Editor re: paper by Wood, Wolferth and Geckeler. Histologic demonstration of accessory muscular connections between auricle and ventricle in the case of short P-R interval and prolonged QRS complex. Appeared in *Amer. Heart J.* **25**: 454-462.
120. ERLANGER, J. 1944. Obituary. Albert Ernest Taussig. 1871-1944. *Tr. Assoc. Amer. Physicians.* **58**: 35-36.
121. ERLANGER, J. 1945. A reassessment of Beaumont the investigator. An address delivered at William Beaumont's grave November 22, 1945. *Weekly Bull. St. Louis Med. Soc.* **40**: 147-150.
122. ERLANGER, J. 1945. Obituary. William Henry Howell. 1860-1945. *Science* n.s. **101**: 575-576.
123. ERLANGER, J. 1945. Obituary. William Henry Howell. 1860-1945. *Year Book of the American Philosphical Society.* Philadelphia: American Philosophical Society. pp. 370-373.
124. ERLANGER, J., and H. S. GASSER. 1945. As funcoes altamente diferenciadas da fibra nervosa. *Resenha. Clin. Client.* **14**: 147-148.

125. ERLANGER, J., and G. M. SCHOEPFLE. 1946. A study of nerve degeneration and regeneration. *Amer. J. Physiol.* **147**: 550-581.
126. ERLANGER, J., and G. M. SCHOEPFLE. 1949. Relation between spike height and polarizing current in single medullated nerve fibers. *Amer. J. Physiol.* **159**: 217-232.
127. ERLANGER, J. 1949. Some observations on the responses of single nerve fibers. Pp. 173-195 in *Les Prix Nobel en 1947*. Imprimerie Royale. Stockholm: P. A. Norstedt & Soner, 1949.
128. ERLANGER, J. 1950. Biographical memoir of William Henry Howell. 1860-1945. *Biographical Memoirs of the National Academy of Science*. **26**: 153-180.
129. ERLANGER, J., and G. M. SCHOEPFLE. 1950. Local responses of single medullated fibers in a nerve. *Amer. J. Physiol.* **163**: 748.
130. ERLANGER, J., and G. M. SCHOEPFLE. 1951. Further observations on local responses of single medullated fibers in frog's phalangeal nerve. *Fed. Proc.* **10**: 120-121.
131. ERLANGER, J., and G. M. SCHOEPFLE. 1951. Observations on the local responses of single medullated nerve fibers. *Amer. J. Physiol.* **167**: 134-146.
132. ERLANGER, J. 1964. A physiologist reminisces. *Ann. Rev. Physiol.* **26**: 1-14.
133. ERLANGER, J. 1968. Joseph Erlanger 1874-1965. *Physiologist* **11**: 1-2.

BIBLIOGRAPHY OF HERBERT SPENCER GASSER

1888–1963

1. BRADLEY, H. C., and H. S. GASSER. 1911-12. Intestinal absorption. (Proc. Am. Soc. Biol. Chem.) *J. Biol. Chem.* **11**: xx.
2. GASSER, H. S., and A. S. LOEVENHART. 1912-13. On the mechanism of stimulation by oxygen want. (Proc. Am. Soc. Biol. Chem.) *J. Biol. Chem.* **14**: xxx.
3. GASSER, H. S., and A. S. LOEVENHART. 1914. The mechanism of stimulation of the medullary centers by decreased oxidation. *J. Pharm. Exptl. Therap.* **5**: 239-273.
4. GASSER, H. S., and W. J. MEEK. 1914. The acceleration of the heart in exercise. (Proc. Am. Physiol. Soc.) *Am. J. Physiol.* **33**: xx.
5. PETERSON, M. S., and H. S. GASSER. 1914. The effect of chemical products of muscular activity on the frequency and force of the heart beat. *Am. J. Physiol.* **33**: 301-312.
6. GASSER, H.S., and W. J. MEEK. 1914. A study of the mechanism by which muscular exercise produces acceleration of the heart. *Am. J. Physiol.* **34**: 48-71.
7. GASSER, H. S. 1917. The significance of prothombin and of free and combined thrombin in blood-serum. *Am. J. Physiol.* **42**: 378-394.
8. ERLANGER, J., R. GESELL, and H. S. GASSER. 1916. The combination of thrombin by the anti-thrombin of the blood-serum. (Proc. Am. Physiol. Soc.) *Am. J. Physiol.* **42**: 608.
9. ERLANGER, J., R. GESELL, H. S. GASSER, and G. L. ELLIOTT. 1917. An experimental study of surgical shock (preliminary report). *J. Am. Med. Assoc.* **69**: 2089-2092.
10. ERLANGER, J., R. GESELL, H. S. GASSER, and G. L. ELLIOTT. 1917-18. Some reactions in the development of shock by diverse methods. *Am .J. Physiol.* **45**: 546.
11. MEEK, W. J., and H. S. GASSER. 1917. A method for the determination of blood volume. (Proc. Am. Physiol. Soc.) *Am. J. Physiol.* **45**: 547.
12. GASSER, H. S., W. J. MEEK, and J. ERLANGER. 1917-18. The blood volume changes in shock and the modification of these by acacia. (Proc. Am. Physiol. Soc.) *Am. J. Physiol.* **45**: 547.

13. Meek, W. J., and H. S. Gasser. 1917-18. The effects of injecting acacia. (Proc. Am. Physiol. Soc.) *Am. J. Physiol.* **45**: 548.

14. Gasser, H. S. 1918. The blood volume changes in shock and methods in which they may be modified. *J. Missouri Med. Assoc.* **15**: 233.

15. Erlanger, J., and H. S. Gasser. 1918. The treatment of standardized shock, I. *Compt. Rend. Soc. Biol.* **81**: 898.

16. Erlanger, J., and H. S. Gasser. 1918. The treatment of standardized shock, II. *Compt. Rend. Soc. Biol.* **81**: 905.

17. Meek, W. J., and H. S. Gasser. 1918. Blood volume. A method for its determination with data for dogs, cats and rabbits. *Am. J. Physiol.* **47**: 302-317.

18. Erlanger, J., R. Gesell, and H. S. Gasser. 1919. Studies in secondary traumatic shock. I. The circulation in shock after abdominal injuries. *Am. J. Physiol.* **49**: 90-116.

19. Erlanger, J., and H. S. Gasser. 1919. Studies in secondary traumatic shock. II. Shock due to mechanical limitation of blood flow. *Am. J. Physiol.* **49**: 151-173.

20. Erlanger, J., and H. S. Gasser. 1919. Studies in secondary traumatic shock. III. Circulatory failure due to adrenalin. *Am. J. Physiol.* **49**: 345-376.

21. Erlanger, J., and H. S. Gasser. 1919. Hypertonic gum acacia and glucose in the treatment of secondary traumatic shock. *Ann. Surg.* **69**: 389-421.

22. Erlanger, J., and H. S. Gasser. 1919. The treatment of standardized shock. *J. Missouri Med. Assoc.* **16**: 98.

23. Gasser, H. S., J. Erlanger, and W. J. Meek. 1919. Studies in secondary traumatic shock. IV. The blood volume changes and the effect of gum acacia on their development. *Am. J. Physiol.* **50**: 31-53.

24. Gasser, H. S., and J. Erlanger. 1919. Studies in secondary traumatic shock. V. Restoration of the plasma volume and of the alkali reserve. *Am. J. Physiol.* **50**: 104-118.

25. Erlanger, J., and H. S. Gasser. 1919. Studies in traumatic shock. VI. Statistical study of the treatment of measured trauma with solutions of gum acacia and crystalloids. *Am. J. Physiol.* **50**: 119-148.

26. Erlanger, J., and H. S. Gasser. 1919. Studies in secondary traumatic shock. VII. Note on the action of hypertonic gum acacia and glucose after hemorrhage. *Am. J. Physiol.* **50**: 149-156.
27. Gasser, H. S., and H. S. Newcomer. 1921. The application of the thermionic vacuum tube to the study of nerve physiology. (79th Meeting Washington Univ. Med. Soc., Dec., 1921.) *J. Missouri Med. Assoc.* **18**: 461.
28. Gasser, H. S., and H. S. Newcomer. 1921. Physiological action currents in the phrenic nerve. An application of the thermionic vacuum tube to nerve physiology. *Am. J. Physiol.* **57**: 1-26.
29. Gasser. H. S., and J. Erlanger. 1922. The cathode ray oscillograph as a means of recording nerve action currents and induction shocks. (Proc. Am. Physiol. Soc., Dec. 1921.) *Am. J. Physiol.* **59**: 473.
30. Gasser, H. S., and J. Erlanger. 1922. A study of the action currents of nerve with the cathode ray oscillograph. *Am. J. Physiol.* **62**: 496-524.
31. Gasser, H. S., and J. Erlanger. 1922. The components of the action currents obtained from nerves. (Proc. Am. Physiol. Soc.) *Am. J. Physiol.* **63**: 417.
32. Leake, C. D., H. S. Gasser, and A. S. Loevenhart. 1923-24. Synergism between sodium cyanide and minimal electrical stimulation of the vagi on the respiratory center. (Proc. Am. Physiol. Soc., Dec., 1923.) *Am. J. Physiol.* **68**: 129.
33. Erlanger, J., and H. S. Gasser, with the collaboration, in some of the experiments, of G. H. Bishop. 1924. The compound nature of the action current of nerve as disclosed by the cathode ray oscillograph. *Am. J. Physiol.* **79**: 624-666.
34. Gasser, H. S., and A. V. Hill. 1924. The dynamics of muscular contraction. *Proc. Roy. Soc. Ser. B.* **96**: 398-437.
35. Gasser, H. S., and W. Hartree. 1924. The inseparability of the mechanical and thermal responses in muscle. *J. Physiol. London.* **58**: 396-404.
36. Gasser, H. S. 1924. The methods of recording the electrical potential change in nerve, with special reference to the use of the Braun tube oscillograph. *Brit. J. Radiol.* (Roentgen Soc. Section.) **20**: 105-111.

37. Erlanger, J., G. H. Bishop, and H. S. Gasser. 1925. Further observations on the analysis of the action current in nerve. (Proc. Am. Physiol. Soc., Dec., 1924.) *Am. J. Physiol.* **72**: 197.
38. Bishop, G. H., J. Erlanger, and H. S. Gasser. 1925. Data on the nature of propagation in nerve. (Proc. Am. Physiol. Soc., Dec., 1924.) *Am. J. Physiol.* **72**: 199.
39. Gasser, H. S., and J. Erlanger. 1925. The nature of conduction of an impulse in the relatively refractory period. *Am. J. Physiol.* **73**: 613-635.
40. Lapicque, L., H. S. Gasser, et A. Desoille. 1925. Relation entre le degré d'hétérogénéité des nerfs et la complexité de leur courant d'action. *Compt. Rend Soc. Biol.* **92**: 9.
41. Erlanger, J., G. H. Bishop, and H. S. Gasser. 1925-26. On conduction of the action potential wave through the dorsal root ganglion. *Proc. Soc. Exptl. Biol. Med.* **23**: 372.
42. Erlanger, J., G. H. Bishop, and H. S. Gasser. 1926. The refractory phase in relation to the action potential of nerve. (Proc. Am. Physiol. Soc., Dec., 1925.) *Am. J. Physiol.* **76**: 203.
43. Bishop, G. H., J. Erlanger, and H. S. Gasser. 1926. The record of the action potential of nerve at the site of stimulation. (Proc. Am. Physiol. Soc., Dec., 1925.) *Am. J. Physiol.* **76**: 204.
44. Erlanger, J., G. H. Bishop, and H. S. Gasser. 1926. Experimental analysis of the simple action potential wave in nerve by the cathode ray oscillograph. *Am. J. Physiol.* **78**: 537-573.
45. Erlanger, J., G. H. Bishop, and H. S. Gasser. 1926. The action potential waves transmitted between the sciatic nerve and its spinal roots. *Am. J. Physiol.* **78**: 574-591.
46. Bishop, G. H., J. Erlanger, and H. S. Gasser. 1926. Distortion of action potentials as recorded from the nerve surface. *Am. J. Physiol.* **78**: 592-609.
47. Gasser, H. S. 1925-26. The response of plexus-free preparations from the small intestine to drugs. (Proc. Am. Soc. Pharmacol. Exptl. Therap., Dec., 1925.) *J. Pharmacol. Exptl. Therap.* **27**: 250.
48. Gasser, H. S. 1926. Plexus-free preparations of the small intestine. A study of their rhythmicity and of their response to drugs. *J. Pharmacol. Exptl. Therap.* **27**: 395-410.

49. GASSER, H. S., and H. H. DALE. 1926. The pharmacology of denervated mammalian muscle. II. Some phenomena of antagonism, and the formation of lactic acid in chemical contracture. *J. Pharmacol. Exptl. Therap.* **28**: 287-315.
50. DALE, H. H., and H. S. GASSER. 1926. The pharmacology of denervated mammalian muscle. I. The nature of the substances producing contracture. *J. Pharmacol. Exptl. Therap.* **29**: 53-67.
51. GASSER, H. S., and J. ERLANGER. 1926-27. The differential action of pressure of fibers of different sizes in a mixed nerve. *Proc. Soc. Exptl. Biol. Med.* **24**: 313-314.
52. GASSER, H. S., and J. ERLANGER. 1927. The role played by the sizes of the constituent fibers of a nerve trunk in determining the form of its action potential wave. *Am. J. Physiol.* **80**: 522-547.
53. ERLANGER, J., H. S. GASSER, and G. H. BISHOP. 1927. The absolutely refractory phase of the alpha, beta and gamma fibers in the sciatic nerve of the frog. *Am. J. Physiol.* **81**: 473.
54. GASSER, H. S. 1927. The relation of the shape of the action potential of nerve conduction velocity. (Proc. Am. Physiol. Soc., April 1927.) *Am. J. Physiol.* **81**: 477.
55. GASSER, H. S. 1928. The relation of the shape of the action of nerve to conduction velocity. *Am. J. Physiol.* **84**: 699.
56. GASSER, H. S. 1928. The recording of single nerve action potentials with the cathode ray oscillograph and the analysis of a volley from the central nervous system into a phrenic nerve. *Am. J. Physiol.* **85**: 372.
57. GASSER, H. S. 1928. The analysis of individual waves in the phrenic electroneurogram. *Am. J. Physiol.* **85**: 569-576.
58. HINSEY, J. C., and H. S. GASSER. 1928-29. The Sherrington phenomenon. *Am. J. Physiol.* **87**: 368-380.
59. GASSER, H. S., and J. ERLANGER. 1929. The action potential in fibers of slow conduction in spinal roots and somatic nerves. *Proc. Soc. Exptl. Biol. Med.* **26**: 647-649.
60. GASSER, H. S., and J. ERLANGER. 1929. The role of fiber size in the establishment of a nerve block by pressure of cocaine. *Am. J. Physiol.* **88**: 581-591.

61. Erlanger, J., and H. S. Gasser. 1929. The action potential in fibers of slow conduction in spinal roots and somatic nerves. (Abstracts of Communications to the Thirteenth International Physiological Congress.) *Am. J. Physiol* **90**: 338.
62. Gasser, H. S., and J. Erlanger. 1929. The method of recording the action potential of nerve with cathode ray oscillograph. (Demonstration.) (Abstracts of Communications to the Thirteenth International Physiological Congress.) *Am. J. Physiol.* **90**: 356.
63. Gasser, H. S. 1929. Arthur S. Loevenhart. *Science.* **70**: 317-321.
64. Erlanger, J., and H. S. Gasser. 1930. The action potential in fibers of slow conduction in spinal roots and somatic nerves. *Am. J. Physiol.* **92**: 43-82.
65. Hinsey, J. C., and H. S. Gasser. 1930. The component of the dorsal root mediating vasodilation and the Sherrington contracture. *Am. J. Physiol.* **92**: 679-689.
66. Gasser, H. S. 1930. Contracture of skeletal muscle. *Physiol. Rev.* **10**: 35-109.
67. Gasser, H. S., and J. Erlanger. 1930. The ending of the axon action potential, and its relation to other events in nerve activity. *Am. J. Physiol.* **94**: 247-277.
68. Graham, H.T., and H. S. Gasser. 1931. Modification of nerve response by veratrine, protoveratrine and aconitine. *J. Pharmacol. Exptl. Therap.* **43**: 163-185.
69. Gasser, H. S. 1931. Nerve activity as modified by temperature changes. *Am. J. Physiol.* **97**: 254-270.
70. Schmitt, F. P., and H. S. Gasser. 1931. The effect of carbon monoxide on the after-potential of medullated nerve. *Am. J. Physiol.* **97**: 558.
71. Gasser, H. S., and H. T. Graham. 1932. The end of the spike-potential of nerve and its relation to the beginning of the after-potential. (Proc. Am. Physiol. Soc., April, 1932.) *Am. J. Physiol.* **101**: 37.
72. Gasser, H. S., and H. T. Graham. 1932. The end of the spike-potential of nerve and its relation to the beginning of the after-potential. *Am. J. Physiol.* **101**: 316-330.

73. Gasser, H. S. 1933. Axon action potentials in nerve. (Cold Spring Harbor Symposium on Quantitative Biology.) *Collecting Net.* **8**: 138.
74. Gasser, H. S., and H. T. Graham. 1933. Potentials produced in the spinal cord by stimulation of dorsal roots. *Am. J. Physiol.* **103**: 303-320.
75. Schmitt, F. O., and H. S. Gasser. 1933. The relation between the after-potential and oxidative processes in medullated nerve. *Am. J. Physiol.* **104**: 320-330.
76. Hughes, J., and H. S. Gasser. 1933. The response of the spinal cord to two stimuli. *Am. J. Physiol.* **105**: 57.
77. Hughes, J., and H. S. Gasser. 1934. Some properties of the cord potentials evoked by a single afferent volley. *Am. J. Physiol.* **108**: 295-306.
78. Hughes, J., and H. S. Gasser. 1934. The response of the spinal cord to two afferent volleys. *Am. J. Physiol.* **108**: 307-321.
79. Gasser, H. S. 1934. Electrical phenomena in nerve. *Occasional Publ. Am. Assoc. Advan. Sci.* **79**: 26-29.
80. Gasser, H. S. 1934. Conduction in nerves in relation to fiber types. *Res. Publ. Assoc. Res. Nervous Mental Disease* **15**: 35-39.
81. Gasser, H. S., and H. T. Graham. 1934. Augmentation of the positive after-potential of nerves by yohimbine. *Proc. Soc. Exptl. Biol. Med.* **32**: 553-556.
82. Gasser, H. S. 1935. Changes in nerve-potentials produced by rapidly repeated stimuli and their relation to the responsiveness of nerve to stimulation. *Am. J. Physiol.* **111**: 35-50.
83. Richards, G. H., and H. S. Gasser. 1935. After-potentials and recovery curve of C fibers. *Am. J. Physiol.* **113**: 108.
84. Clark, D., J. Hughes, and H. S. Gasser. 1935. Afferent function in the group of nerve fibers of slowest conduction velocity. *Am. J. Physiol.* **114**: 69-76.
85. Gasser, H. S. 1935. The changes produced in the injury discharge of nerve fibers during the period following a propagated disturbance. (Proc. Physiol. Soc., July, 1935.) *J. Physiol. London* **85**: 15P.

86. GASSER, H. S. 1936. Potentials developing in mammalian nerve fibers during activity and their relation to excitability. Pp. 317-326 in "The Problems of Nervous Physiology and Behavior," dedicated (Jubilee Symposium) to Professor J. S. Beritoff under the auspices of Georgian Branch Acad. Sci. USSR. Tiflis, Georgia, USSR, 1936.
87. GASSER, H. S., and H. GRUNDFEST. 1936. Action and excitability in mammalian A fibers. *Am. J. Physiol.* **117**: 113-133.
88. GASSER, H. S. 1936-37. The control of excitation in the nervous system. *Harvey Lectures.* 169-193.
89. GASSER, H. S. 1937. Reciprocal innervation, pp. 212-218 in "Jubilee Volume in honor of Professor J. Demoor," Thone, Liege.
90. ERLANGER, J., and H. S. GASSER. 1937. Electrical signs of nervous activity. Eldridge Reeves Johnson Foundation for Medical Physics. Philadelphia: Univ. of Penna. Press. 221 p.
91. GASSER, H. S. 1938. Electrical signs of biological activity. *J. Appl. Physics* **9**: 88-96.
92. GASSER, H. S. 1938. Recruitment of nerve fibers. *Am. J. Physiol.* **121**: 193-202.
93. GASSER, H. S., D. H. RICHARDS, and H. GRUNDFEST. 1938. Properties of the nerve fibers of slowest conduction in the frog. *Am. J. Physiol.* **123**: 299-306.
94. GRUNDFEST, H., and H. S. GASSER. 1938. Properties of mammalian nerve fibers of slowest conduction. *Am. J. Physiol.* **123**: 307-318.
95. GASSER, H. S., and H. GRUNDFEST. 1939. Axon diameters in relation to the spike dimensions and the conduction velocity in mammalian A fibers. *Am. J. Physiol.* **127**: 393-414.
96. GASSER, H. S. 1939. Axons as samples of nervous tissue. *J. Neurophysiol.* **2**: 361-369.
97. GASSER, H. S. 1940. Methods of analysis of nervous action, pp. 250-257 in "Chemistry and Medicine; Papers Presented at the Fifteenth Anniversary of the Founding of the Medical School of the University of Minnesota," M. B. Visscher (ed.). Minneapolis, University of Minnesota Press.

98. GASSER, H. S. 1941. The classification of nerve fibers. *Ohio J. Sci.* **41**: 145-159.
99. GASSER, H. S. 1943. Pain producing impulses in peripheral nerves. *Res. Publ. Assoc. Res. Nervous Mental Disease* **23**: 44-62.
100. GASSER, H. S. 1944. The Nobel Prizes. *Science* **100** (Suppl. 10) Nov. 10, 1944.
101. ERLANGER, J., and H. S. GASSER. 1945. As funcoes altamente diferenciadas da fibra nervosa. *Resenha Clin. Cient.* **14**: 147.
102. GASSER, H. S. 1945. Mammalian nerve fibers. Nobel Lecture, December 12, 1945, pp. 128-141 in "Les Prix Nobel en 1940-1944." Stockholm, Nobelstiftelsen.
103. GASSER, H. S. 1946. Simon Flexner 1863-1946. *Trans. Assoc. Am. Physicians* **59**: 11-13.
104. GASSER, H. S. 1947. Protocol for a review of psychiatry. Psychiatric research. Pp. 69-79 in "Papers Read at the Dedication of the Laboratory for Biochemical Research, McLean Hospital, Waverly, Massachusetts, May 17, 1946." Cambridge, Mass., Harvard Univ. Press.
105. GASSER, H. S. 1950. Unmedullated fibers originating in dorsal root ganglia. *J. Gen. Physiol.* **33**: 651-690.
106. GASSER, H. S. 1954. Acceptance of the Kober medal for 1954. *Trans Assoc. Am. Physicians* **67**: 44-46.
107. GASSER, H. S. 1954. Conduction in unmedullated fibers in dorsal root. *Am. J. Physiol.* **179**: 637.
108. GASSER, H. S. 1954-55. Unmedullated nerve fibers. *Trabajos Inst. Cajal Invest. Biol.* **45**: 1-8.
109. GASSER, H. S. 1955. Properties of dorsal root unmedullated fibers on the two sides of the ganglion. *J. Gen. Physiol.* **38**: 709-728.
110. GASSER, H. S. 1955. Sir Henry Dale: his influence on science. *Brit. Med. J.* **1**: 1359-1360.
111. GASSER, H. S. 1956. Olfactory nerve fibers. *J. Gen. Physiol.* **39**: 473-496.
112. GASSER, H. S. 1956. Fibras nerviosas amielinicas con funcion aferents. *Anales Inst. Farmacol. Espan.* **5**: 11-26.
113. GASSER, H. S. 1957. Alan Gregg (1890-1957), pp. 126-129 in "Year Book of the American Philosophical Society."

114. GASSER, H. S. 1958. The postspike positivity of unmedullated fibers of dorsal root origin. *J. Gen. Physiol.* **41**: 613-632.
115. GASSER, H. S. 1958. Comparison of the structure, as revealed with the electron microscope, and the physiology of the unmedullated fibers in the skin nerves and in the olfactory nerves. *Exptl. Cell Res. Suppl.* **5**: 3-17.
116. GASSER, H. S. 1960. Effect of the method of leading on the recording the nerve fiber spectrum. *J. Gen. Physiol.* **43**: 927-940.
117. GASSER, H. S. 1964. Herbert Spencer Gasser, prepared by Joseph C. Hinsey. *Experimental Neurology*, Supplement **1**. Academic Press. viii + 38pp.

INDEX

INDEX

Lightning Source UK Ltd.
Milton Keynes UK
UKHW022139210722
406200UK00003B/321